Satellites as they cross the night sky look like moving stars, which can be accurately tracked by an observer with binoculars as well as by giant radars and large cameras. These observations help to determine the satellite's orbit, which is sensitive to the drag of the upper atmosphere and to any irregularities in the Earth's gravity field. Analysis of the orbit can be used to evaluate the density of the upper atmosphere and to define the shape of the Earth.

Desmond King-Hele was the pioneer of this technique of orbit analysis, and this book tells us how the research began, before the launch of Sputnik in 1957. For thirty years King-Hele and his colleagues at the Royal Aircraft Establishment, Farnborough, developed and applied the technique to reveal much about the Earth and air at a very modest cost. In the 1960s the upper-atmosphere density was thoroughly mapped out at heights of 100 to 2000 km, revealing immense variation of density with solar activity and between day and night. In the 1970s and 1980s a picture of the upper-atmosphere winds emerged, and the profile of the pear-shaped Earth was accurately charted. The number of satellites now orbiting the Earth is over 5000.

This book is the story of how this orbital research developed to yield a rich harvest of knowledge about the Earth and its atmosphere, in a scientific narrative that is enlivened with many personal experiences.

A TAPESTRY OF ORBITS

A TAPESTRY OF ORBITS

DESMOND KING-HELE

CAMBRIDGE
UNIVERSITY PRESS

CAMBRIDGE UNIVERSITY PRESS
Cambridge, New York, Melbourne, Madrid, Cape Town, Singapore, São Paulo

Cambridge University Press
The Edinburgh Building, Cambridge CB2 2RU, UK

Published in the United States of America by Cambridge University Press, New York

www.cambridge.org
Information on this title: www.cambridge.org/9780521393232

First published 1992
This digitally printed first paperback version 2005

A catalogue record for this publication is available from the British Library

Library of Congress Cataloguing in Publication data
King-Hele, Desmond
A tapestry of orbits/Desmond King-Hele.
p.   cm.
Includes bibliographical references and index.
ISBN 0 521 39323 X
1. Artificial satellites – Orbits.   I. Title.
TL 1080.K55 1992
629.43´4 – dc20   92-11674   CIP

ISBN-13 978-0-521-39323-2 hardback
ISBN-10 0-521-39323-X hardback

ISBN-13 978-0-521-01732-9 paperback
ISBN-10 0-521-01732-7 paperback

Contents

Preface

This book is a personal account of the researches based on analysis of satellite orbits between 1957 and 1990 at the Royal Aircraft Establishment, Farnborough, work in which I played a leading role. The book is most definitely not an impartial history of the subject world-wide: contributions by other groups are mentioned only when necessary. Nor is the book an autobiography, though the science is punctuated – and perhaps enlivened – by some personal experiences.

A book of this kind, a hybrid of science and life, presents the author with many stylistic problems. I have ruthlessly gouged out as many 'I's as possible, and have tried to avoid mentioning too many names (with apologies to all those who find themselves liquidated). I decided to use 'we' quite often: throughout the book *we* means 'those of us at the RAE who were concerned with or working on the problem'. Individual names are mentioned too, of course, and often the *we* is defined by giving the authors of a paper in a note.

I have tried to make the book widely intelligible to readers without specialized knowledge. There is a light sprinkling of mathematical equations: but if you don't like them you can skip them without losing the thread.

Most spacecraft chatter continuously, sending back to the ground stations so much data that storage can be quite a problem. The satellites selected for orbit analysis, on the other hand, are usually dumb (and deaf and blind): but they can be seen from the ground as they cross the sky, and from the observations their orbits can be determined. The changes in the orbits can be measured to obtain detailed information about the Earth's upper atmosphere and gravity field (from which the shape of the Earth can be derived). Thus small changes in satellite orbits lead to great changes in perceptions of the Earth and air. Acclaimed as the most cost-effective form of space research in the 1960s, orbit analysis lost some of its shine in the 1970s, when wider horizons beckoned space scientists as expensive 'hitech' spacecraft came into vogue. By the 1980s orbit analysis was in a siding rather than on the main line of space research: but the techniques were improved and the work continued to produce new and relevant results, as the book demonstrates.

Prologue, 1948–1953

The morning was sunny and serene, the day was Monday 12 September 1948, and I was travelling by train to begin a new life working at the Royal Aircraft Establishment at Farnborough in Hampshire. As the steam-engine puffed along the last few miles from Guildford to the curiously-named North Camp station, I had no idea what was in store, never having ventured into Hampshire before (unnecessary travel had been frowned on during the Second World War). During the previous two years I had been working for a mathematics degree at Cambridge, and it was in the garden of the Cambridge Appointments Board in May that I was interviewed by two 'Men from the Ministry' and offered a post in the Guided Weapons Department at the RAE, as an alternative to three years of military service. My interviewers were very pleasant and persuasive, and the alternative was also persuasive: I accepted the post as a temporary Scientific Officer at the excellent salary of £340 a year, though with various deductions.

At first sight, the Royal Aircraft Establishment created a favourable impression, because I had seen nothing like it before. It covered about three square miles and seemed like a small town. Some of the buildings were rather scruffy, but some were quite presentable, and the built-up area was balanced by the extensive airfield. There were about 10000 people working at the RAE then, and the whole place seemed to be buzzing with activity, the noisiest buzzing being produced by the frequent take-offs and landings of jet aircraft.

My second impression was not so good, for the rest of the day was spent learning the first lesson of bureaucracy, that individuals wait while the system creaks on. There was at least something to see at the Personnel Department, on the second floor of the main building, overlooking the airfield: the first Farnborough Air Show had finished on the previous day, and all the aeroplanes were departing. The really long wait, of two hours

1

in a prefab hut, was for the Medical Officer. When at last the great man came, he blithely dismissed me without examination, saying 'you look quite healthy'. The next hurdle was the news that I should not be working at the RAE after all, but at an outstation, Bramshot Golf Club, three miles to the west. This proved to be a country house that had spawned a cluster of prefabs: the golf course was defunct, but the view was still attractively rural. Bramshot also lived up to its name of 'outstation': it had its own main-line railway station, optimistically called Bramshot Halt, though in reality the trains to Bournemouth and Exeter all went through at high speed along the straight and level 15 miles between Farnborough and Basingstoke.

The atmosphere of the Bramshot office was quite relaxed, as was the Head of the Guided Weapons (GW) Department, Ronald Smelt, who soon afterwards went to the USA and later became Vice-President of Lockheeds. I was assigned to the Assessment Division, headed by W. H. Stephens, one of my Cambridge interviewers. As Stephens was away, I went on to see Clifford Cornford, the second interviewer: he sometimes greeted new recruits sitting with his feet on the mantelpiece, but this time he was the right way up. Most guided missiles, he said, had rocket engines, but now there was a new idea in the offing, missiles powered by ramjets. Soon, working in a section headed by C. L. Barham, I was deep in a report entitled 'The gas dynamic theory of the athodyd', learning the mysteries of supersonic aerodynamics and combustion thermodynamics.

There was a bus to Bramshot each day at 8.20 a.m., and the journey soon became routine, apart from the daily race against time by one of the youngest travellers, Doreen Gilmore, who came by train to North Camp station and then cycled the remaining two miles, usually arriving just in time. As it turned out, she was to work closely with me for many years.

In March 1949 we moved from Bramshot into the RAE, to 134 Building (later called Q134), and there was a new Head of GW Department, Morien Morgan, a live-wire Welshman who gave the impression that anything was possible (and was later the main driving-force behind Concorde). Q134 was solidly built in the 1930s, and simply designed, with an east–west corridor about 200 yards long on each of three floors, and offices or labs with high ceilings and large windows, facing north or south. My office was to be there for the next 39 years. For the first four of those years, the Head of Assessment Division was the sprightly Bill Stephens, notable for his mid-Atlantic accent and outsize American car. Clifford Cornford was the dynamo, keeping everyone alert and keen, and taking his part in the hand-computing too, to encourage the computers (in those days they were human). During these years in C. L. Barham's section my task was to make sketch designs and performance estimates for missiles powered by ramjet or rocket, to meet the new 'operational requirements' continually being conjured up by the Army, Navy or Air Force. The RAE, as part of the Ministry of Supply, was there to supply the answers. I was certainly 'on tap not on top', and this missile assessment work continued until 1957. By then I had beavered away on missile designs for seventeen different projects, every one of which was subsequently cancelled.

In retrospect this work seems boring and useless, but it was not like that at the time. What a surprise it was to be paid for mathematical researches into unknown territory, and provided with quite a pleasant environment! There were other brightnesses too in the early 1950s. I met Marie Newman, who was to become my wife in 1954, and my early fascination with the poetry of Shelley (especially its scientific facets) was leading me towards writing a book about him. More frivolously, there was the thrill of being able to buy a box of chocolates for the first time when the long dark night of rationing came to an end. Other hardships of the War years were also fading from memory, as was the trauma of Cambridge in January–March 1947, when the temperature never rose above freezing point, my only heating was a coal fire with a meagre ration of coal, and Trinity College impounded my bread units, only to give much of the bread to the dons (an injustice that biased me against life in Cambridge). The missile work was looking up too, and I was able to work on a theoretical paper about the stability of asymmetrical missiles (RAE Report GW 16). The theory explained, in terms of roll–yaw resonance, why a current ramjet test vehicle sometimes became disastrously unstable. Above all, there was now a chance of studying satellite orbits, thanks to the discerning and forward-looking attitude of the three who directed my work, Morgan, Stephens and Cornford. They saw that the Space Age was imminent; and I had the vague feeling that research on orbits might offer me a satisfying career.

Though this prologue is out of keeping with the rest of the book, it introduces the people who most influenced me and helps to explain the background from which the later space research developed.

1

Prelude to space, 1953–1957

And gazing burns with unallow'd desires.
                    Erasmus Darwin, *The Loves of the Plants*

It was in 1953 that the metamorphosis of missiles into satellites began. One important new start was the prospect of rockets for upper-atmosphere research. The impetus came from a group of scientists belonging to the Royal Society's Gassiot Committee, particularly Professor Harrie Massey of University College London, and Professor David Bates of the Queen's University, Belfast. The existence of the Gassiot Committee was an extraordinary stroke of luck for space science, as I came to realize much later. The Royal Society covers all science, and until 1935 the one exception to this rule was the Gassiot Committee, the Society's only specialized 'in-house' committee: it had been formed in 1871, to oversee Kew Observatory, and was expanded during the Second World War to cover atmospheric physics in general. The Gassiot Committee was vitally important for two reasons: first, it was a preconstructed official pathway into space; second, the Royal Society was fully committed from the outset, thus making respectable a subject dismissed by many as 'utter bilge'.

The Gassiot Committee organized an Anglo-American conference on rocket exploration of the upper atmosphere, at Oxford in August 1953, and this can now be seen as the first British step on the ladder into space which we climbed for nearly twenty years. I cannot remember much about the meeting, except that it was held in a dark medieval lecture-room, lit by a few light bulbs with dusty white shades: it seemed paradoxical that these new ventures into space were being planned in such antiquated surroundings.

Having listened intently to the American researchers speaking at Oxford, the scientists of the Gassiot Committee were keen to have a British rocket

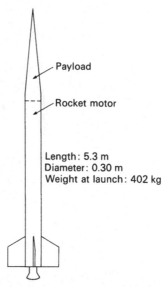

Payload

Rocket motor

Length: 5.3 m
Diameter: 0.30 m
Weight at launch: 402 kg

Fig. 1.1. Sketch design of upper-atmosphere rocket. Reproduced from RAE Technical Note GW 315 (1954).

to explore the upper atmosphere: a few months later they came to visit the Director of the RAE, Sir Arnold Hall. To my surprise, I was called in, because they wanted a design study for a rocket to fly to high altitudes, and I was the obvious victim, having already done so many (abortive) design studies. But this new idea looked as though it might actually materialize. My design study was issued in May 1954 as RAE Technical Note GW 315: it showed that a solid-fuel rocket to go up 50 miles should be feasible, and there were sketches of possible layouts: Fig. 1.1 shows the favoured one. The Deputy Director of the RAE, Dr F. E. Jones, was keenly supporting the project and, after fuller design studies by Derek Dawton and others, it came to fruition as the Skylark rocket (Fig. 1.2). Aerodynamically, the main design problem was to keep the centre of pressure as far to the rear as possible with the minimum of tail fin: that is why the early Skylark had swept-back fins, and three fins instead of four. Skylark was about 50% larger than the sketch design in linear dimensions, and consequently 3 times as heavy.

The RAE looked after the Skylark project with great success for many years, thanks largely to Frank Hazell and Eric Dorling. Some 200 Skylarks were launched in the UK programme between 1957 and 1978, many of them reaching heights of more than 200 km. Their scientific instrumentation led on to the scientific payloads of the six Ariel satellites

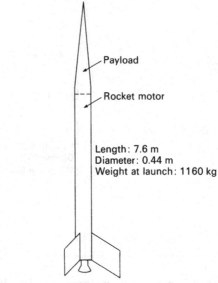

Fig. 1.2. Early Skylark rocket (1957).

launched between 1962 and 1979. Both the Skylark and the Ariel projects are described in considerable detail in the excellent *History of British Space Science* by Harrie Massey and M. O. Robins. I was not involved in either programme after the initial design study, so there are only passing references to Skylark and the Ariels in the rest of this book.

Long-range ballistic rockets

The second important new start in 1953 was to prove more fruitful for me. Early in the year we heard about long-range ballistic rockets being developed by the USSR. Previously, ballistic missiles had been un-mentionable; but by the end of the year they had flipped over to respectability, and for the next four years Doreen Gilmore and I spent a substantial part of our time producing a series of lengthy reports specifying the performance of long-range ballistic missiles. The first report, RAE Technical Note GW 305 (March 1954), dealt with the trajectories *in vacuo*, which I always saw as satellite orbits that didn't quite 'make it', falling back to Earth because their velocities were less than the satellite speed (about 7.9 km/s). An example of the orbital bias was the method of specifying the 'optimum' trajectory (minimum velocity for a given range) not by the optimum climb angle at all-burnt, but by the optimum

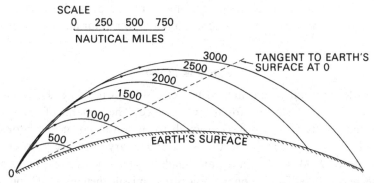

Fig. 1.3. Optimum trajectories for ballistic missiles. Reproduced from RAE Technical Note GW 305 (1954). The numbers on the curves indicate the impact ranges in nautical miles. The optimum trajectory has an initial climb angle that gives maximum impact range.

eccentricity of the 'orbit', namely $\tan\left(\dfrac{\pi}{4}-\dfrac{\text{range}}{4R}\right)$, where R is the Earth's radius. Fig. 1.3 shows the optimum trajectories to scale, the ranges being given in nautical miles, our standard unit at that time ($= 1.853$ km).

After this orbital frolic came a serious series of reports running to about 100 000 words, with 300 pages of detailed diagrams covering all aspects of performance – the effects of structure weight, rocket specific impulse, climb path, the number of stages of propulsion, and so on.[1] The detailed calculations were done by Doreen Gilmore, whose accuracy, speed and efficiency set new standards for me. All the work on ballistic missiles was of course hand-calculation with electro-mechanical calculators, Friedens and Monroes, which took about half a minute to grind out a division. Every two or three years, a faster and quieter machine arrived, and eventually there was a machine that could take square roots on its own: before that, the quickest way to find the square root of a number was to divide it by a guessed value and halve the difference, and then do the same again, if need be. With these slow calculators, the approximations in the mathematical analysis needed to be both bold and of wide validity if the results were to be of any use. It was good practice for the future satellite orbit analysis.

In 1954 there were changes in the RAE hierarchy: Morien Morgan became Deputy Director on the Aircraft side, with F. E. Jones continuing as the other Deputy Director, covering the GW Department, which was now headed by W. H. Stephens, with Clifford Cornford in charge of the Assessment Division.

Our work on ballistic missiles continued into 1955: satellites remained stuck somewhere over the horizon, too hot a potato for anyone to grasp.

Satellites at last

'Lift-off' eventually came in the summer of 1955, when F. E. Jones asked for a study of a satellite for reconnaissance. The autumn of 1955 was devoted chiefly to this project (though work continued on the abortive missile designs). In our report,[2] issued in January 1956, we proposed a satellite of 2000 pounds mass, and made design studies for a two-stage launcher, with half an eye on the Blue Streak missile then in its early phases of development. Like Blue Streak, the proposed satellite launcher relied on liquid oxygen and kerosene as propellants for the first stage. We tried to work out a near-optimum climb path, and the chosen trajectory was quite similar to those subsequently used by real satellites. The weight of the launch vehicle came out as 60 tons, and it was rather similar to the later US Thor-Delta 1 launcher.

For the reconnaissance we selected an orbit inclined at 60° to the equator, so that the land masses up to latitudes of 60° N (or a little more) could be covered. The orbit needed to be as low as possible for good photography, but not so low as to be brought down too quickly by air drag. The chosen orbit was circular at a height of 200 nautical miles (370 km), with an orbital period of 91.8 minutes. The satellite's lifetime in this orbit was estimated at about 100 days, enough to complete the reconnaissance, though the lifetime could be extended by using auxiliary rocket motors to return to a high orbit.

Our 370 km orbit was quite close to those later chosen for the early Soviet photographic reconnaissance satellites beginning in 1962: their orbits were inclined at 65° to the equator, near-circular, and at heights near 300 km, with orbital periods near 90 minutes. Their heights were lower because they only had to stay in orbit for about a week. Hundreds of these Soviet reconnaissance satellites, each of several tons mass, have been launched in the past thirty years: the orbital inclinations have ranged between 62° and 82°, and the heights usually between 200 and 400 km. Who would have thought in 1956 that, thirty years later, the USSR would be selling high-quality photographs of Britain, taken from space?

An interesting novelty in our 1956 satellite report was a set of maps of the satellite tracks over the Earth for various inclinations to the equator, 60°, 80° and 90°. The map for 60° inclination, shown as Fig. 1.4, proved very useful when Sputnik 1 was launched in October 1957 into an orbit at

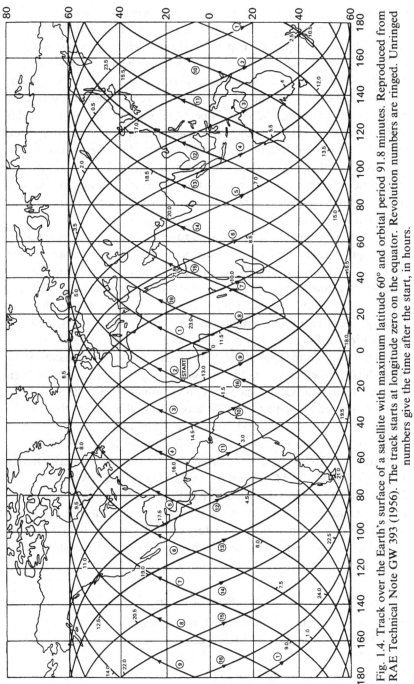

Fig. 1.4. Track over the Earth's surface of a satellite with maximum latitude 60° and orbital period 91.8 minutes. Reproduced from RAE Technical Note GW 393 (1956). The track starts at longitude zero on the equator. Revolution numbers are ringed. Unringed numbers give the time after the start, in hours.

65° inclination with not too different an orbital period (95 minutes initially, but quickly decreasing). The 1956 report also showed how the reconnaissance could be completed strip by strip and, after allowing for poor weather, we suggested 100 days for completion. The photographs were to be transmitted back to Earth each day by radio, thus avoiding the problem of the fierce heating that the satellite would have to endure if it was to be brought back to Earth. The skin temperature of the satellite when in orbit was estimated as between 220 K (for white paint) and 350 K (for black paint).† This seemed to present no problem. Nor did meteor hazards. We also discussed the guidance accuracy for orbit injection; the requirements for attitude control; auxiliary rockets for counteracting drag; and possible power supplies, including a radioactive heat source, or the new-fangled 'solar battery'.

Most of the recommendations have proved to be valid and practicable, and the project then outlined, far from being outdated, still smacks of the futuristic. The idea of Britain putting into orbit a 1 ton satellite with a home-grown launcher seems further away now than it did then. Today it is only other countries, like China, Japan, India and Israel, that can manage such feats; but confidence was high in those days, and the cancellation of Blue Streak was not foreseen.

For the moment the light of good fortune still shone. The Director of the RAE during the crucial years 1955–59 was George Gardner, who had been the first Head of the GW Department on its formation in 1946. Sir George Gardner (he was knighted in 1959) was very well-liked as Director: he seemed to know everyone, and often dropped in unannounced for a chat about the work and the world. In 1956 F. E. Jones left the RAE to join Mullards. W. H. Stephens was promoted to succeed him as Deputy Director, with Clifford Cornford becoming Head of the GW Department. Further down the line, I was promoted to Principal Scientific Officer. It was also fortunate that the report on the reconnaissance satellite attracted the attention of Sir Owen Wansbrough-Jones, Chief Scientist of the Ministry: he sent a very kind letter and would obviously be a supporter of work on space.

The effects of air drag

Two questions were left unanswered in the report on the reconnaissance satellite. First, what was the effect of air drag on the orbit? And that included the satellite's plunge through the lower atmosphere at the end of

† Temperatures in kelvins, that is, 273 plus degrees centigrade. Thus 220 K is −53 °C, and 350 K is 77 °C.

its life. These problems were tackled in a further long report, again in collaboration with Doreen Gilmore, issued in September 1956 as RAE Technical Note GW 430 and entitled 'The descent of an earth satellite through the atmosphere'. We assumed that the Earth and atmosphere were spherical, and considered initially-circular orbits only. Down to heights well below 200 km, we found, the satellite descends in a spiral at a speed equal to the circular orbital velocity at the current height – about 7.8 km/s at a height of 200 km.

This result was independent of the mass, size and shape of the satellite, and showed that the satellite was *not* slowed by air drag, but slightly increased its speed as it descended. The angle of descent, in radians, turned out to be twice the drag/weight ('weight' being the mass multiplied by the acceleration due to gravity at the current height). This shows why the speed increases: the satellite adjusts its angle of descent so that its acceleration due to descending is twice the deceleration due to drag. This simple result did not seem to be in any papers or books in the previous century, and we thought it new. Many years passed before I found that the spiral descent path had been derived by Sir Isaac Newton in the *Principia* in 1687. His results are in a somewhat different form, but essentially he did obtain an equivalent solution, as we explained thirty years later[3] (during which time no one else pointed out the equivalence).

From our supposedly-new theory, plus an assumed model of air density ρ versus height, we could calculate the lifetimes of satellites of various area/mass ratios, and the graph given in 1956 is reproduced as Fig. 1.5. The numbers on the curves are the values of the area/mass parameter $\Delta = SC_D/m$, where S is the satellite's cross-sectional area (in square feet here), m is its mass (in pounds) and C_D its drag coefficient,† taken at that time as 2.0. (Multiply Δ by 0.2 to convert to m^2/kg.) Our warning that the model of density versus height 'will no doubt prove to be grossly inaccurate when the true values are known' was needed, but at the relevant heights near 200 nautical miles the error was not so great as in some later models (we had used the model in the 1950 US Handbook of supersonic aerodynamics). The lifetime of the reconnaissance satellite in the actual atmosphere of 1957 would have been about 200 days, rather than the 100 days we estimated. In years of low solar activity, such as 1964 or 1985, the lifetime would have been much longer.

† The drag coefficient C_D is related to the drag D by the equation $D = \frac{1}{2}\rho v^2 SC_D$, where v is the satellite's velocity: the value of C_D depends on the shape, height and angular motion of the satellite and is not yet accurately calculable; for most satellites at heights of 200–400 km, C_D is near 2.2.

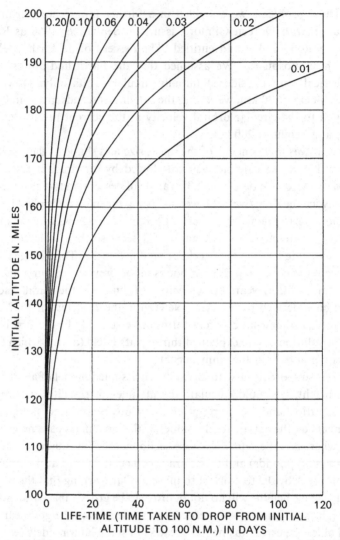

Fig. 1.5. Lifetime of a satellite in a circular orbit. Reproduced from RAE Technical Note GW 430 (1956). Numbers on the curves indicate the values of $\Delta = SC_D/m$. See text for explanation of symbols.

For heights below 200 km, our simple result for the descent angle became less accurate, with errors of 2% at 170 km: so we had to devise approximations for the velocity and descent angle as power series in Δ. With the aid of these approximations, and the somewhat erroneous atmospheric model, the descent path could be calculated, for various values of Δ, down to a height of about 60 km. With the 'standard' reconnaissance satellite of the 1956 report, which had Δ = 0.002 square feet/pound, the calculated angle of descent increased from 0.005° at a height of 110 nautical miles (203 km) to 0.025° at 90 nautical miles (167 km).

At heights below 40 nautical miles (73 km) we changed over to numerical integration of the descent path. For the standard satellite the angle of descent increased from 2.5° at 30 nautical miles (55 km) to 5.9° at 10 nautical miles (18 km) and 73° at sea level. The speed stayed above 6 km/s down to a height of 25 km, and then quickly decreased, with the greatest deceleration, 12 g, at heights near 15 km. The horizontal distance covered, from a height of 60 nautical miles (110 km) was 4400 km, and the descent from 40 nautical miles (73 km) took just over 4 minutes, assuming that the satellite survived the intense heating, which was greatest at heights near 15 km. If the skin were of steel half-an-inch thick, we calculated the mean skin temperature as 2000 K; so a greater thickness or a more refractory material would be needed to save the satellite from breaking up.

Blue Streak and Black Knight

Any danger of the brain overheating in these torrid calculations was averted by returning to the cold climate of the seventeen abortive design studies. The sixteenth of these was more interesting than most. The report on the reconnaissance satellite had assumed a British launcher: would that be possible? In particular, what would happen if Black Knight was put on top of Blue Streak to try to make a satellite launcher? The ballistic missile Blue Streak, by now well-advanced in development, had a weight at launch of about 90 tons, a length of 62 feet and a diameter of 10 feet. Black Knight was a smaller rocket designed for testing the high-speed atmospheric entry of Blue Streak's warhead. Black Knight weighed about 6 tons, was 30 feet long and 3 feet in diameter. (Later, the first stage of the Black Arrow satellite launcher was built round the propulsion systems of two Black Knights.)

The proposed design for a satellite launcher[4] had Blue Streak as the first stage, Black Knight as the second, and a third stage of total mass 3000 pounds including a small integral rocket motor for the final insertion into orbit. We calculated the total impulse required from the third-stage rocket motor, and hence its weight: a payload of 2400 pounds could be placed in a circular orbit at a height of 200 nautical miles or 2200 pounds if the height

was 400 nautical miles. The weakest feature of the work was in estimating
– it was little more than guessing – the structure weight for the staging,
strengthening and separation. But the study served to show the feasibility
of having Blue Streak as a first stage for a satellite launcher. Later, after
being cancelled as a missile, Blue Streak did become the first stage of the
satellite launcher developed in the 1960s by ELDO (the European
Launcher Development Organization). In the many test firings of the
ELDO launcher, Blue Streak always worked well, but one of the upper
stages, designed and made in France and Germany, always failed. In the
end the launcher was abandoned, and Ariane was started. So, after many
years of false promise, this study also fell into the pit where the other
sixteen abortive missile designs could be seen wriggling feebly.

Satellite orbits

Good luck was still the order of the day, because demand from the military
for mundane missile assessment eased off in the spring of 1957, giving the
chance for an attack on the second unsolved question of satellite motion:
'How is the orbit affected if the Earth is oblate rather than spherical?' We
approached the problem naively, without any skill in the arcane arts of
'celestial mechanics' and never having heard of 'Lagrange's planetary
equations'. This naivety proved to be a great advantage, because we did
not have to hack away a jungle of wrong preconceptions. Instead, starting
from the basic equations of motion, we were able to proceed to a solution
by bold assumptions and brute force.

   This is a useful moment to pause and look at a diagram of a satellite orbit
about the Earth, Fig. 1.6, that defines the terms used. The *perigee P* is the
point nearest to the Earth, and the *apogee A* the point most distant. The
major axis of the ellipse is the length AP and the *semi major axis*, denoted
by a, is a half of AP. The shape of the ellipse is measured by the *eccentricity*,
denoted by e, which is equal to the difference between the apogee and
perigee height, divided by the major axis; and e is always less than 1 for an
ellipse. For a circular orbit, the perigee and apogee heights are the same,
and so $e = 0$. Most of the orbits appearing later in the book have
eccentricities between 0 and 0.3. The orbit shown in Fig. 1.6 has an
eccentricity near 0.7, because the diagram is clearer when e is large. Fig. 1.6
also shows that the distance of the perigee from the Earth's centre is
$a(1-e)$ and the distance of the apogee from the Earth's centre is $a(1+e)$.
The satellite S, travelling anticlockwise, has its position specified by the

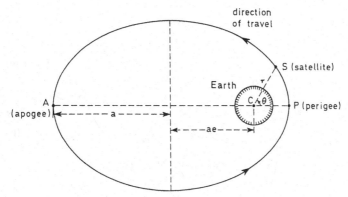

Fig. 1.6. Elliptic satellite orbit.

angle $P\hat{C}S$, marked as θ in Fig. 1.6. The distance r of the satellite S from the Earth's centre C at any point on the elliptic orbit is then given by

$$r = \frac{a(1-e^2)}{1+e\cos\theta}.$$

This is all that needs to be known about the geometry of the ellipse in the rest of this book: it is mainly the semi major axis a and the eccentricity e that arise frequently; but if you have this equation for r as well, you can find your way around the ellipse.

Fig. 1.6 shows how the satellite moves in its own orbital plane; but another diagram is needed to define the angular position of the orbital plane relative to the Earth, and this is the purpose of Fig. 1.7. The Earth is taken as a sphere, with the north pole at the top and the equator round the middle, as usual. Part of the satellite orbit, including P and S from Fig. 1.6, is shown on the right, and the plane of the orbit is taken as 'slicing through' the Earth, the unbroken curve through N being the visible cut, and the broken curve being the cut on the invisible side.

Two quantities are needed to specify the angular position of the orbit relative to the Earth. The first is the angle between the orbit and the equator, known as the *orbital inclination* and written as i. The inclination has already sidled into my discussion of the reconnaissance satellite, for which the orbital inclination was 60°. The maximum latitude of the satellite track over the Earth is virtually equal to the inclination (exactly so for a spherical Earth, and very nearly so for the real oblate Earth).

The second quantity for specifying the angular position of the satellite's orbital plane is some measure of the longitude. The geographical longitude is unhelpful, because it changes too fast: the Earth spins at about 15° per

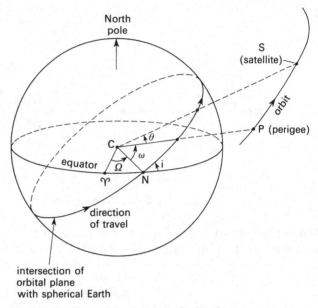

Fig. 1.7. Satellite orbit relative to the Earth.

hour, while the satellite plane remains nearly fixed in space. Instead, some
fixed direction in the sky is needed to act as a marker: the point usually
chosen is 'the first point of Aries', denoted by the peculiar symbol ♈,
originally intended to represent the horns of a ram.† The longitude of the
orbital plane is then defined by the angle, measured along the equator,
between ♈ and the point *N*, called the 'ascending node', where the satellite
crosses the equator going north. This angle is denoted by the Greek symbol
Ω (capital omega), as shown in Fig. 1.7: the full name of Ω is 'right
ascension of the ascending node', but I shall call it the *longitude of the node*.
If the Earth were spherical, Ω would remain virtually constant; with the
real, oblate Earth, Ω changes slowly and steadily – and accurate measurement of the changes reveals a great deal about the detailed shape of the
Earth.

These and other scientific results about the Earth and its atmosphere are
the main theme of this book, and most of the results have come from
looking at the changes in the four quantities already defined, namely the

---

† Strictly, ♈ is the point where the Sun crosses the plane of the Earth's equator at the spring
equinox, and ♈ does move slowly in the sky (at 0.014° per year) because the Earth's axis
gyrates round the heavens once every 26000 years. But this snailpace movement is usually
too slow to worry about. Nor does it matter that ♈ has now moved out of the constellation
Aries into Pisces and should have exchanged its horns for something fishy.

semi major axis *a*, the eccentricity *e*, the inclination *i* and the longitude of the node Ω.

However, there is one further parameter needed, to specify the angular position of the perigee point relative to the equator. This is shown in Fig. 1.7 as the angle ω (small omega), which has a rather strange name, the *argument of perigee* – there is no argument about it, you just take it or leave it. When the perigee is at the northward crossing of the equator (the point *N*), ω is zero. When the perigee is at the maximum latitude north, $\omega = 90°$; when the perigee is at the southward equator crossing, $\omega = 180°$; when it is at maximum latitude south, $\omega = 270°$. For a highly eccentric orbit, the latitude of perigee is important, because it controls the height of the satellite over a particular area. But it is less important for a near-circular orbit and of no importance at all for an exactly circular orbit.

Learning about these five parameters *a*, *e*, *i*, Ω and ω – called *orbital elements* – is the initiation ceremony for understanding the real-life behaviour of the actual satellites that will figure in later chapters of this book. It is learning the orbital alphabet, beginning with *a* and ending with omega, but much easier because only vowels appear, and in the right order (omega being the Greek for o when pronounced as in *old*).

Orbits about the oblate Earth

After this digression on orbital elements, I can return to the work early in 1957 with Doreen Gilmore on 'The effect of the Earth's oblateness on the orbit of a near satellite', to quote the title of the completed report, which was issued as RAE Technical Note GW 475 in October 1957. The analysis was successful largely because of an arbitrary – but, as it turned out, very fruitful – assumption about orders of magnitude. The gravitational pull on a satellite near the Earth is altered by up to 1 part in about 500 by the effects of oblateness, that is, by the effect of the 'second harmonic' in the gravitational field, which is associated with a numerical coefficient *J* having a value near 0.002.† The theory was easiest for orbits of small eccentricity. So we assumed that *e* was less than 0.05, and called this 'first order'. The effects of oblateness, and of e^2, were 'second order', 1 in 500, which fitted in verbally with 'second harmonic'. 'Third order' was 1 in 10000, and 'fourth order' was 1 in 200000, which was (conveniently) believed to be the order of magnitude of the fourth harmonic in the geopotential.

The solution for the radial distance was carried to the third order, with fourth-order results only for very small values of *e*. Not knowing any

† *J* is equal to $1.5 J_2$, where J_2 is defined by the equation on p. 40.

better, we worked in terms of the angular travel ψ in the orbital plane starting from the apex, the point of maximum latitude north. (After three decades 'in the wilderness' ψ was resurrected in 1988 as the most efficient angular variable.)[5] If the Earth were spherical, the orbit would be an ellipse, with the radial distance r given by the equation on p. 15. When the satellite is moving in the gravitational field appropriate for an oblate Earth, the radial distance r would, we hoped, be expressible in terms of an ellipse which was rotating in its own plane at a constant rate, though there would also be small changes in r, of order J and Je (second and third order). On these assumptions we managed to find quite simple solutions, showing that the distortion of the gravitational pull caused by the Earth's oblateness led to an oscillation in the radial distance twice per revolution, with amplitude $0.94(R/a)\sin^2 i$ nautical miles (though we used the symbol α rather than i, and r rather than a). As neither R/a nor $\sin^2 i$ can exceed 1, these changes in orbital height are less than 1 nautical mile. This nicely justified the original assumption of a rotating ellipse as a good first approximation to the real orbit, thus legitimizing the method.

Though the changes in radial distance are only small, and oscillatory, two other important orbital effects build up cumulatively. First, as already mentioned, the orbital plane rotates. To quote the original report:

The orbital plane, instead of remaining fixed, rotates about the Earth's axis in the opposite direction to the satellite, at a rate of $10.0 \ (R/\bar{r})^{3.5} \cos \alpha$ degrees per day, where α is the inclination of the orbital plane to the equator, R the Earth's equatorial radius, and $\bar{r}$ the satellite's mean distance from the Earth's centre.

If terms of order e^2 are neglected, $\bar{r} = a$. The above expression for the rate of rotation due to J has proved to be correct, though it can now be written more accurately, with e^2 terms added and i for the inclination, as $\dot{\Omega} = -9.964(R/a)^{3.5}(1-e^2)^{-2} \cos i$ degrees per day. The negative sign means that the orbital plane swings westwards for a satellite launched towards the east, with i less than 90° (as most satellites are, to take advantage of the Earth's rotation). The precession of the orbital plane for satellites of oblate planets was previously known, but had not been given explicitly for the Earth in this way.

The second important finding concerned the rotation of the ellipse within the orbital plane. To quote again from the 1957 report:

The major axis of the orbit rotates in the orbital plane at a rate of $5.0(R/\bar{r})^{3.5}(5\cos^2 \alpha - 1)$ degrees per day. Thus, for 200 n.miles orbital altitude, it rotates at about 16 degrees per day in the same direction as the satellite for a near-equatorial orbit, and at about 4 degrees per day in the opposite direction for a polar orbit. There is no rotation when $\alpha = 63.4°$.

In 1957 this result came as rather a surprise and caused some controversy; but it has stood the test of time. The more accurate current version, with e^2 terms added, is $\dot{\omega} = 4.982\,(R/a)^{3.5}(1-e^2)^{-2}(5\cos^2 i - 1)$ degrees per day. Strangely enough, the 'critical inclination' (as it is now called) of 63.4° does not seem to have been publicized previously, though I feel sure that the $(5\cos^2 i - 1)$ variation must appear in some eighteenth- or nineteenth-century papers or books on celestial mechanics. However, I have not yet located such a source.

What I did find a few years later was a paper in German by H. G. L. Krause, presented at the Seventh International Astronautical Congress in 1956 (and published in 1957), covering some of the same ground as we did, and showing that the perigee rotates at a rate proportional to $(5\cos^2 i - 1)$. However, this result was buried in the mathematics, with no indication that it was important; so it is not surprising that no one told us. If I had attended the Congress and had known of this paper, would I have thought it worth while to start on our 'pioneering' theory, which was the key opening the door to satellite orbit analysis? The answer is probably 'yes', but the approach might have been different.

In 1957 our theory seemed very much a 'shot in the dark' by Earthbound authors unqualified in the ancient mysteries of celestial mechanics. However, two reassuring tests of the theory were possible. The first was an internal comparison with a more accurate theory. Our main results were obtained by 'perturbation methods', neglecting small terms of order Je^2. For equatorial orbits, however, we could solve the problem for any value of e, exact to order J and ignoring only J^2 terms. It was good to find that, for equatorial orbits, the results from the perturbation theory agreed with the more exact solution. This comparison was rather incestuous, but the second test was independent. Although no real satellite had been launched in the summer of 1957, a numerical integration of one particular orbit had been performed in 1956 by G. Fosdick and M. Hewitt. There was good agreement between our results and their numerical integration, after correcting a 23% error in their value of J. Our results for the radial distance r were also subsequently confirmed.

This report on the effect of the Earth's oblateness on orbits was completed by September 1957. It was a good start towards understanding the behaviour of a real satellite in an elliptic orbit of small eccentricity. The orbital plane would rotate from east to west at a rate of $10(R/a)^{3.5}\cos i$ degrees per day; the perigee point would rotate in the orbital plane at a rate of $5(R/a)^{3.5}(5\cos^2 i - 1)$ degrees per day; and the radial distance r would still be, within about 1 km, as for an unperturbed ellipse. We knew that air drag acting on a circular orbit produced a spiral descent path, the descent angle being twice the drag/weight ratio. The effect of drag on an elliptic orbit was still known only qualitatively: the eccentricity would decrease as the life went on, but the exact form of the decrease was not known, and was the next problem waiting to be tackled.

At that moment, on 4 October 1957, Sputnik 1 appeared in the sky, sweeping scientists out of the ivory tower into the real-life maelstrom of the Space Age. It was sink or swim, and, thanks to the theoretical lifebelts, we had a better chance than those who plunged in unprepared.

2

The real thing, 1957–1958

Like stars to their appointed height they climb.

P. B. Shelley, *Adonais*

The launch of Sputnik 1 on 4 October 1957 was a traumatic event for the USA and much of the western world. For years there had been an unspoken assumption that the Russians were dark and backward people, and that all new initiatives in science and technology occurred, almost as a natural law, in 'the West'. Disbelief was widespread. 'What I say is truth, and truth is what I say', that popular saying of the 1980s, had its adherents in the 1950s too, and they assured the world that Sputnik 1 was just propaganda and was not really in orbit at all.

My view of the event was different. For several years we had been showing in theory how ballistic rockets could be turned into satellite launchers by adding a small upper stage to produce the necessary extra velocity. The USSR had launched an intercontinental rocket in August 1957, and little extra velocity would be needed to attain orbit. So it would be quite easy for the USSR to launch a small satellite like Sputnik 1, which was a sphere 58 cm in diameter of mass 84 kg with four long aerials (Fig. 2.1). The real surprise was the final-stage rocket that accompanied Sputnik 1 into orbit. The rocket appeared much brighter than the pole star as it crossed the night sky, and seemed likely to be at least 20 m long, far larger than anything contemplated in our paper-studies of satellites: the final-stage rocket for our reconnaissance satellite was less than 5 m long.

Launching such a bright rocket was an unplanned master-stroke by the USSR. People could see the rocket crossing the sky, and came out to look in their thousands. Most of them thought they were seeing the satellite itself, but this error scarcely mattered. Even the most sceptical seemed

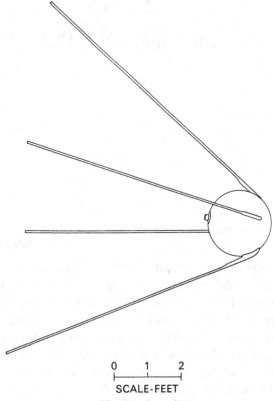

0   1   2
├─────┴─────┤
SCALE-FEET
Fig. 2.1. Sputnik 1.

unwilling to deny the evidence of their own eyes, and had to admit there was *something* going over. The brightness of the rocket was also beneficial (from a selfish viewpoint) in arousing enthusiasm for accurate optical observing of satellites, from which reliable orbits could be computed. If the early satellites had been faint, people would have become blasé by the time the bright ones appeared, and the enthusiasm would not have arisen. As it was, the researches that are the subject of this book were set going with a momentum that endured for many years.

When William Wordsworth looked back to the time of the French Revolution, which had seemed to promise a new era, he wrote:

> Bliss was it in that dawn to be alive,
> But to be young was very heaven!

That rather overstates the effect of the satellite launchings, but it was exhilarating to have worked for four years on the theory of satellite launching and orbital motion, and then suddenly to find a scholar's fantasy turn into reality. Poets down the ages have imagined space travel, but they never saw it in reality, as I was lucky enough to do. As a mathematician, I was fascinated to see Newton's laws in daily action, with the satellite appearing so regularly each night, yet always slightly earlier than it would have been *in vacuo*, because of the drag of the tenuous upper atmosphere. Newton's laws were only half the story, however: I had been interested in astronomy since reading Sir James Jeans's *The Stars in their Courses* as a teenager, but the static tableau of the night sky induced a touch of boredom. These bright moving lights visible to the naked eye were much more interesting, and it was not long before I was lured into observing the satellites visually – and continued doing so, because the observations proved to be so helpful in determining the orbits, which could then be analysed for research purposes. So the launching of Sputnik 1 created two different challenging pursuits: first, the mathematical orbit analysis, the research I pursued for the rest of my scientific career; and, second, the exacting coordination of brain, eye and hand that goes into the art of satellite observing.

The timing of Sputnik 1 also worked out well for me. If the launch had been a year earlier, the theories would not have been ready. A year later the theories would have been widely known and we should not have had a 'flying start'. The timing also suited me for a more selfish reason. By 1957 my book about Shelley's poetry was virtually finished: the subsequent furore would otherwise have delayed its completion by several years. Not being qualified in literature, I was very lucky to get the book published by Macmillan, and a few years later that luck would probably have gone away.

Four days after the launch of Sputnik 1, there was a Press conference in London that I remember for two curious reasons. First, it was my one-and-only journey from the RAE to London in an official chauffeur-driven car (it was all downhill after that – train and tube). The other reason is not so silly. The Royal Society had agreed to hold the Press conference in their rooms at Burlington House: for me it was a memorable first visit there, to be followed by many more, and a long association with the Society. There were four representatives of the Ministry of Supply: the Chief Scientist Sir Owen Wansbrough-Jones, W. H. Stephens, Clifford Cornford and myself. It was not long before a journalist asked, 'Could the Russians drop a bomb on us from this satellite as it passes over?' My senior colleagues decided to pass this awkward query to me for an answer. As I remember, I said that if the Russians wished to do this it would be much easier to start from the ground rather than from orbit; and that a bomb 'dropped' from the satellite would merely go round in orbit alongside it. Was I being let out of my institution into public for the first time just to calm the nerves of the nation? Looking back afterwards, I wondered whether the question had been 'planted'; if so, I had no prior knowledge of it. Certainly it was better to have the question answered by a naive young scientist than by an Establishment figure pre-programmed to say 'there is no cause for alarm'.

The tracking of Sputnik 1

Sputnik 1 was famous for its 'bleep', produced by a powerful radio transmitter operating at 20.005 MHz and 40.002 MHz, and powered by chemical batteries. Indeed this was virtually the only instrumentation of Sputnik 1, and very effective it was in announcing the new era of space flight.

These strong signals gave radio scientists the chance to take the lead in tracking the satellite. The first and most obvious technique is just to listen in and measure the change in frequency due to the Doppler effect as the satellite crosses the sky. When a satellite rises above the horizon, the signals come in at a frequency about 1 kHz higher than the standard, then there is a rapid decrease to the standard as the satellite passes closest, and a further decrease to 1 kHz below the standard as it recedes into the distance. It is like standing on a railway station as a whistling train passes through.

Recording the variation of frequency with time, as shown in Fig. 2.2, easily tells you two things. You can find the time of closest approach to the receiving station, when the frequency is at the standard value; and you can also estimate the satellite's distance of closest approach as it passes by – the closer the approach, the more rapid the change in frequency. By making such measurements at two or more stations at least 100 km apart, you can determine the time, height and track of the satellite. For Sputnik 1 the results were not expected to be very accurate, because the radio signals were of a frequency low enough to suffer appreciable distortion by the ionosphere, so much so that the simple picture of Fig. 2.2 was rarely recorded. Much more accurate Doppler tracking is possible at higher frequencies and later formed the basis of the US Navy's Navigation Satellite System.

A second method of radio tracking is the interferometer, a rather long name for a quite simple arrangement. Basically, you need to have two aerials at the same height above the ground, separated by a known

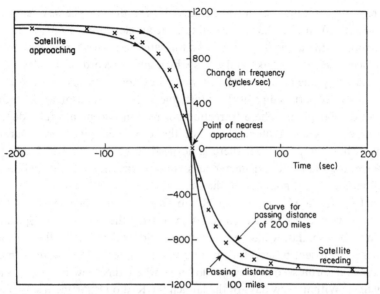

Fig. 2.2. The change in the frequency of radio signals when a satellite passes, as a result of the Doppler effect. The nearer the satellite passes, the more quickly the change occurs: so, from the observed values (the line of crosses), the passing distance can be estimated (here 160 miles). Reproduced from *Satellites and Scientific Research* (1960).

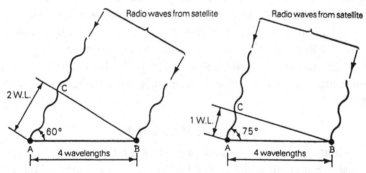

Fig. 2.3. Principle of the radio interferometer. The two aerials A and B are sited 4 wavelengths apart. When the phase difference between the waves arriving at the two aerials is equivalent to 2 wavelengths (left), the satellite is at an elevation of 60°. When it is 1 wavelength (right), the elevation is 75° (or 75.52°, to be precise).

distance, perhaps 4 wavelengths as in Fig. 2.3. Then the signals from the aerials are brought together and compared. Whenever the signals are the same, 'in phase', the distance of the satellite from the two aerials must differ by an exact number of wavelengths. If the difference is 2 wavelengths, as on the left in Fig. 2.3, the satellite is at an elevation of 60°; if it is 1 wavelength, the angle is near 75°. As the satellite passes across the sky it produces a series of peaks and zeros which give a record of the elevation angle in the plane of the aerials – assume they are on an east–west line. A second pair of aerials on a north–south line gives the elevation angle in the perpendicular plane. These 'direction cosines' in two planes fully define the direction of the satellite. In practice, the zero signals give more accurate results than the peaks, but there is no need to go into further detail. In essence, the radio interferometer tells you the direction of the satellite at each moment as it goes across the sky.

In 1957 the RAE had a large Radio Department, where many of the scientists were eager to be involved in measuring the signals from Sputnik 1. To start with, they wanted to see if they could track it, but for them it was also a heaven-sent beacon for probing the ionosphere. They were keen to establish its orbit so that they would know where their new beacon was at any time. Within a few days of the launch the Radio Department set up an interferometer at Lasham airfield, a quiet site 12 miles southwest of Farnborough. The signals used were those at 40 MHz (7.5 m wavelength). There were two pairs of dipole aerials at right angles, the aerials in each pair being about 30 m apart, or 4 wavelengths. This hastily-constructed instrument worked very well and often detected the satellite on transits when it rose only a few degrees above the horizon. On some nights up to eight transits were recorded. The angular accuracy was estimated to be about 0.1°. A rather similar interferometer was constructed by scientists at the Mullard Radio Astronomy Observatory at Cambridge.

Doppler measurements were also made by the RAE and many other organizations, and gave further information on the height and track of Sputnik 1.

The orbit of Sputnik 1

These initiatives on radio tracking were excellent, but how were the results to be used to determine the orbit? And we needed the orbit in days, not months. By another slice of luck in timing, a digital computer called Pegasus had recently been installed at the RAE. It could make 500 multiplications per second, about 5000 times faster than our existing unprogrammable hand-calculating machines, and was under the command of Robin Merson, who was most skilled in the new art of computing. He at once set to work on devising a program for determining the orbit of Sputnik 1 from the radio interferometer observations.

The determination of the orbit is best approached via the orbital elements a, e, i, Ω and ω defined in the previous chapter. An accurate value of the semi major axis a could be found at once from Kepler's third law, which relates the square of the orbital period T to the cube of the semi major axis a. Algebraically, $T = 2\pi(a^3/\mu)^{\frac{1}{2}}$, where μ is the gravitational constant for the Earth, 398 600 km^3/s^2. If T is measured in *minutes*, as is customary, a factor of 60 needs to be inserted and the numerical equation for a is

$$a = 331.25 T^{\frac{2}{3}} \text{ km.}$$

The interferometer gave the time when the satellite crossed a chosen latitude correct to about 1 s, so the average value of T over one day could be found correct to 1 part in 50 000. (T is calculated from the time between two transits about a day apart, divided by the number of revolutions.) Hence the average value of a over one day would be accurate to 0.1 km, a good start for the orbit determination.

The orbital inclination was difficult to determine from observations at only one latitude (51° N) and the values obtained had a spread of 0.3° about a central value of 65.0°, which was the inclination announced by the USSR. So i was taken as 65.0°, though it subsequently turned out that 65.1° would have been better.

The other orbital parameters (e, Ω and ω) were estimated by a process of adjustment. A plausible set of parameters was chosen, and the computer calculated a succession of values for the direction cosines that would be recorded by the interferometer. These were compared with the real observational values, and the input parameters were adjusted until the two sets agreed as closely as possible.

The Head of the GW Department, Clifford Cornford, set up a small group to work on the determination and analysis of Sputnik 1. The group was led by George Burt and the chief contributors to the work, in alphabetical order, were Doreen Gilmore, myself, David Leslie, Bert Longden and Robin Merson. My main interest was in applying and extending the theory developed in the previous two years.

What did the orbit tell us?

Thanks to the excellent observations and rapid orbit determination, a good orbit of Sputnik 1 was available within about two weeks of its launch.

The first question was, 'Would the orbit help to reveal the shape of the Earth?' The theory for the effect of the Earth's oblateness on the orbit was useful in the orbit determination. We knew that the longitude of the node should decrease by about 3.2° each day, and this could be built into the program. The observational results confirmed this value in an approximate manner, but did not continue long enough to provide a proper test, because the batteries aboard Sputnik 1 ran down and transmissions ceased by 25 October. The perigee should, according to the theory, have moved backwards round the orbit at a rate of about 0.4° per day. Again this was consistent with what was observed, but the observational value was not nearly accurate enough to check the theory. So the answer to the first question was, 'Not quite'.

A more promising avenue for research was to determine the density of the upper atmosphere from the rate at which the orbit of Sputnik 1 was contracting. The orbital period at noon on 15 October was 95 minutes 48.5 seconds, decreasing at 2.28 seconds per day. This information and all the other orbital data appeared in an article written in late October and published in *Nature*[1] on 9 November 1957. To explain how the air density was estimated (and also avoid the danger of covering flaws with a veneer of hindsight) I shall quote from the article:

The air density has been evaluated roughly, on the assumption that the change in major axis is due to a reduction in apogee only, arising from an impulsive change in satellite velocity at perigee. This change in velocity is then compared with the integrated effect of the drag/mass ratio, assuming some particular atmospheric model. The factor of difference between the two values for the change in velocity gives an indication of the factor by which the density in the assumed model is in error. At the satellite altitude, free-molecular flow prevails and, since the isothermal Mach number of the satellite is high, the drag coefficient is taken as 2. This very approximate method gives the density at 130 n.miles altitude as about 2×10^{-10} of that at sea level.

In the calculation Sputnik 1 was taken as a sphere 58 cm in diameter with a mass of 83.6 kg. As sea-level density is 1.2 kg/m³, the density derived here from Sputnik 1 was 2.4×10^{-10} kg/m³ at a height of 241 km (130 nautical miles).

The four crucial assumptions in the method described were at that time all guesses, but fortunately they have proved to be valid, to the order of accuracy needed. First, 95 % of the change in semi major axis *was* due to reduction in the apogee height. Secondly, the variation of density with height was recognized as important and was represented by the best existing model, the ARDC 1956 model,[2] which was not specified at the time because it was unpublished. Thirdly, it was appreciated that the density being registered was not just that at the perigee, but was averaged over a height band with its lower limit at the perigee: the height quoted for the density determination was 11 km above the perigee. Fourthly, the choice of drag coefficient was good: a value near 2.2 would now be recommended.

The value of density at a height of 241 km in the ARDC 1956 model was 2.4×10^{-11} kg/m^3. So Sputnik 1 indicated that the actual density in October 1957 was about 10 times greater than the model. In 1983 I recalculated the density as obtained from the orbit of Sputnik 1, using current theory and methods, and came up with a value of 2.25×10^{-10} kg/m^3 at a height of 247 km, which is equivalent to 2.5×10^{-10} kg/m^3 at 241 km, that is, only 5% different from the value of 2.4×10^{-10} kg/m^3 originally calculated.

However, the value does need a further correction because in 1957 the four aerials attached to Sputnik 1 were ignored, and these probably increased the cross-sectional area by about one third. This reduces the value of density calculated with hindsight to 1.7×10^{-10} kg/m^3 at 247 km, with a possible error of $\pm 10\%$ due chiefly to the errors in calculating the cross-sectional area. Thus the density in October 1957 was really 8 times greater than in the ARDC model.

You might think that these results would now be only of academic interest, and superseded by later measurements. Strangely enough, this is not so, because upper-atmosphere density is strongly dependent on solar activity, and in October 1957 solar activity reached a peak higher than in any previous month in the past 300 years or any subsequent month. Thus there are no other comparable results available. The values from Sputnik 1 remain unique; consequently they deserve some further comment.

The density determined from the orbit of Sputnik 1 was the first of thousands of such measurements made in the 1960s and 1970s. These data served as the basis for various 'reference atmospheres', of which the most widely used has been the COSPAR International Reference Atmosphere 1972, usually abbreviated to *CIRA 1972*. This model[3] gives the variation of density with height for a series of values of the exospheric temperature (i.e. above 500 km). This depends on the values of seven geophysical parameters – namely the current and 3-monthly solar activity indices, the geomagnetic index, the day of the year, the latitude, season and local time. For Sputnik 1 the appropriate value of exospheric temperature proves to be 1500 K, and Fig. 2.4 shows the variation of density with height as given by *CIRA 1972* for temperatures of 1500 and 1800 K. Also shown are the RAE 1957 value, this value as revised in 1983 (sphere only) and the 'corrected 1983 value' with sphere plus aerials, shown with 'error bars'.

The RAE 'corrected 1983' value in Fig. 2.4 suggests that the 1500 K curve is too low, and that a temperature of 1800 K would be nearer the

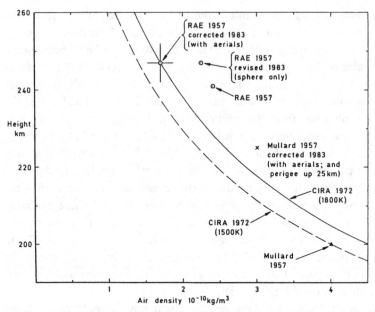

Fig. 2.4. Values of air density obtained from Sputnik 1 in 1957; values as revised in 1983; and curves from *CIRA 1972* (the COSPAR International Reference Atmosphere 1972).

mark. Such an error in *CIRA 1972* is quite possible, because the model is based on data from the 1960s when solar activity was much lower: to find a value for October 1957 requires extrapolation beyond its range of validity.

So far I have given only the RAE results; but the Mullard Radio Astronomy Observatory at Cambridge also tracked Sputnik 1, and derived an orbit published in *Nature* on 2 November 1957. Their orbit was for midnight on 14–15 October, 12 hours earlier than ours. On converting the RAE orbit to this date, the two orbits are very similar. The Mullard orbital period, 95 minutes 50.0 seconds, was only 0.4 seconds greater than ours, and the rates of decay were close, 2.2 and 2.3 seconds per day respectively. Their perigee height, 197 ± 10 km, was much lower than ours, 230 ± 13 km, and not so near the value later determined in retrospect, 224 ± 5 km.

The Cambridge group made an approximate estimate of air density, as 4×10^{-10} kg/m³ at a height of 200 km. Since the method was very approximate and the height quoted was 25 km lower than ever reached by the satellite, this value cannot be taken too seriously. But, by a strange chance, it is exactly on the *CIRA 1972* curve for 1500 K. When this value was revised in 1983, by taking the perigee 25 km higher and taking account of the aerials, it came out 10 % higher than the CIRA curve for 1800 K. If we assume the latter is correct, the error in the Mullard method would have been only 10 %.

Thus the values of density from Sputnik 1, calculated so hurriedly, survive the critical scrutiny of more cynical eyes looking in from an era when the glory of the dawn has faded into the light of common day.

The final-stage rocket of Sputnik 1

The empty rocket that accompanied Sputnik 1 into orbit was in some ways more important than the satellite itself. To begin with, as already mentioned, it was much larger, about 26 m long and 3 m in diameter as we now know, and much heavier, with a mass of about 4 tons. Being so large, the rocket was bright enough to be seen easily with the naked eye as it crossed the night sky, and was also detected in daylight by the new radio telescope at Jodrell Bank when used as a radar.

There was of course no plan for observing such an object: no one expected anything so bright. In the USA the 'Moonwatch' teams were being set up, with the aim of observing the very faint proposed US satellite: a line of visual observers with telescopes would create a 'fence' across the sky. But neither fence nor telescopes were needed with the rocket of Sputnik 1. It was a good moment for that genius for improvisation which in those days Britain was famous for (the building of the interferometers was one example).

The optical-observing improvisation relied largely on the initiative of the Nautical Almanac Office (NAO) at the Royal Greenwich Observatory, Herstmonceux. The NAO was directed by Donald Sadler, and several of the staff, particularly Gordon Taylor, were experienced in timing the occultations of stars by the Moon. From mid October to the end of the year the NAO acted as the UK prediction service for satellites. Armed with the predictions, anyone who wanted to observe the rocket would know when and where to look. The observations sent in would give the time of crossing a particular latitude, and hence the current orbital period, from which predictions for subsequent days could be made. During October the visual observers – a self-appointed group of people keen to observe – came to agree on how the satellites should be observed in a usefully accurate manner. The essential equipment was a stopwatch and good eyes or, even better, binoculars. The observer watched the rocket until it passed between two known stars, pressed the stopwatch as it did so, and also estimated the fractional distance between the two stars, for example, 3/10 up from star

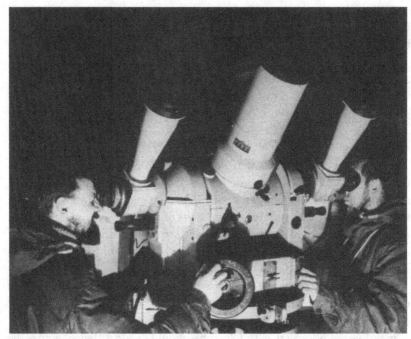

Fig. 2.5. A kinetheodolite in action. This shows the instrument used at the Royal Observatory Edinburgh in the late 1960s: the observers are B. McInnes (left) and R. Eberst (right).

A to star B. In these early days people hoped for accuracies of about 0.2° in direction and 0.5 s in time, after stopping the stopwatch against standard time signals. Ideally, the observations would have been used to determine the orbit, and not just to produce new predictions. However, the rocket could only be seen in Britain on southbound transits during October, and a few observations confined to a small arc are not enough to determine a good orbit. The totality of observations world-wide was needed; but there was no organization to collect these and, with the launch of the even brighter Sputnik 2 early in November, the rocket of Sputnik 1 took second place. For prediction purposes it was assumed that the rocket began life in an orbit similar to that of Sputnik 1, though the rocket's orbit contracted more rapidly because its mass/area ratio was lower and air drag therefore had more effect.

The rocket of Sputnik 1 made its final plunge through the atmosphere on 1 December, after having remained in orbit for 57.6 days. Sputnik 1 itself decayed on 4 January 1958, after a lifetime of 92 days.

It may seem that little was achieved with the rocket of Sputnik 1.

Outwardly that is true; but below the surface its effect was profound, and nowhere more so than in the Trials Department at the RAE, which had great skill in tracking missiles optically with kinetheodolites at its trials ranges, at Aberporth in Wales, Larkhill on Salisbury Plain, Orfordness in Essex and West Freugh in Scotland. If missiles, why not satellites, which appear to move more slowly and certainly move more smoothly?

A kinetheodolite (Fig. 2.5) consists of a telescope cradled in an accurate and heavy mounting, which allows the instrument to be pointed in any direction. It can be rotated horizontally so as to look towards any point of the compass, and elevated to any angle above the horizon. The compass point (azimuth) and the elevation angle are continuously displayed inside the instrument. Two observers control the direction of the telescope, each looking through a smaller side-telescope: one controls the azimuth, the other the elevation, and together they keep the main telescope pointing at the missile (or satellite). Meanwhile flash photographs are being taken at intervals within the kinetheodolite, recording (a) the azimuth and elevation scales, (b) the time, and (c) the image of the object being tracked, showing accurately its position relative to the cross-wires of the telescope.

On the initiative of P. Nuttall-Smith, several kinetheodolites belonging to the Trials Department were modified in the hope of recording fifty or more observations of a bright satellite as it crossed the sky. The expected accuracy in direction was about 0.01°, and this would allow accurate orbit determination. The first kinetheodolites were ready for use at the end of October.

Sputnik 2 arrives

Sputnik 2 was launched at about 4.30 a.m. on 3 November 1957. I was born at about 4.30 a.m. on 3 November 1927. A curious coincidence: but I am not one of those who still believe in supernatural influences, so I felt no frisson. What I did feel that morning is recorded in my poem 'Thirty', in imitation of Dylan Thomas's 'Poem in October':

> It was my thirtieth year to heaven
> Woke to my hearing from over the Alton road
> And the muddy pooled and the seagull
> Sprinkled valley
> The morning beckon
> With thunder growling and the cats and dogs of rain
> And the knock of hailstones on the steamy windows
> Myself to go out
> That second
> Into the noisiness of the still sleeping town.

The poem doesn't mention Sputnik 2, being written before I knew of the launch. The coincidence of dates did have the effect of hardening my habitual scepticism by making me doubt the mathematical theory of probability when the probability is very low. Since then I have always treated any probability of less than 1 in a million as meaningless.

Sputnik 2 was even more of a shock than Sputnik 1, because it carried a dog, Laika, in a cabin. This mini-spaceship was more than 2 m long and 1 m in diameter, with a mass of 508 kg. What was more, Laika survived the launch and so became the first space-traveller. As if all this was not enough, Laika's cabin was still attached to the final-stage rocket, which was similar to the rocket of Sputnik 1. So Sputnik 2 was huge: at the time it was thought to be about 75 feet long and 10 feet in diameter, about the size of a large single-decker bus. This has proved to be an underestimate, as the length was 31.8 m (104 feet) and the maximum diameter 2.95 m (9.7 feet). The total mass of Sputnik 2 was about $4\frac{1}{2}$ tons, again rather greater than was thought at the time.

Sputnik 2 was not only a sensation for the newspapers, it was also another heaven-sent gift for the trackers, who were now 'experienced' (or so they thought). Sputnik 2 was transmitting radio signals and was also extremely bright, often brighter than any star in the sky. So it offered the chance of a combined operation with radio, radar and optical tracking. For its first week in orbit, it was 'all systems go' on the ground in Britain. The radio interferometer and Doppler, the kinetheodolites and visual observers, and the Jodrell Bank radio telescope in radar mode – all were hard at work.

Unfortunately the radio transmitter of Sputnik 2 ceased operation after one week, so we lost the advantage of being able to record many transits per day. However, the kinetheodolite observations were proving to be very accurate, so there was hope of determining a much better orbit than for Sputnik 1. Also the predictions could be kept going very well on the basis of visual observations, aided by some from Jodrell Bank.

The prediction service at the Royal Greenwich Observatory, Herstmonceux, soon ran into a serious obstacle in the form of Dr Richard Woolley, the Astronomer Royal, who controlled the RGO, and was not at all keen on space. Consequently, the service was transferred to the RAE at the beginning of 1958: I was in charge, while Doreen Gilmore produced the detailed predictions. Woolley's hostility to space was another stroke of luck, because it meant that the RAE received all the UK visual and radar observations: the duty of predicting brought with it the need to monitor the orbital period of Sputnik 2 and the privilege of analysing its variations.

The decision to transfer the prediction service to the RAE was triggered by the launch of Sputnik 2, which seemed likely to herald many further launches: more than 3000 launches later, it can safely be said that this idea was right. Being brighter and longer-lived, Sputnik 2 took the limelight away from the rocket of Sputnik 1. Orbit determination and analysis were concentrated on the new object, which was well observed from the start. Its perigee height proved to be much the same as that of Sputnik 1, but the apogee height was much greater (1660 km) and its lifetime of 162 days was consequently longer than the 92 days of Sputnik 1, though not by as much as might be expected, because Sputnik 2 had a lower mass/area ratio than Sputnik 1. By analysing the changes in the orbit of Sputnik 2 as time went on, we were able to open up three new lines of research – on the variations in upper-atmosphere density, on the Earth's gravitational field and on upper-atmosphere winds.

Variations in upper-atmosphere density revealed by Sputnik 2

In 1958 little was known about the shape, size or mass of Sputnik 2; so we could not determine the upper-atmosphere density directly, as with Sputnik 1. Again theory came to the rescue, however, by showing how the orbital decay rate ought to increase as time went on, if the density was constant. Comparing this with the observed decay rate revealed the variations of density with time. The 1956 theory for the effect of air drag on orbits applied only for circular orbits, but the launch of Sputnik 1 showed the need for a theory covering elliptic orbits. It was obvious that the drag suffered by the satellite at perigee would be much greater than at apogee. So the satellite would lose a little speed at each passage of perigee and would not swing out so far at the next apogee. The orbit would become more circular, with the perigee height decreasing only slightly and the apogee height decreasing much more rapidly, as shown in Fig. 2.6. But what was the exact form for the decrease in orbital period and eccentricity with time throughout the life? And how could you estimate the lifetime from the observed initial decay rate?

In collaboration with David Leslie, I began working on these theoretical problems and, despite the hectic ambience, we soon had some answers to the questions. We took the upper atmosphere as spherically symmetrical, with the density ρ decreasing exponentially with distance r from the Earth's centre, so that ρ was proportional to $\exp(-r/H)$, where the quantity H, called the *density scale height*, is assumed constant. The meaning of H is that if the height (or distance r) increases by H, the density

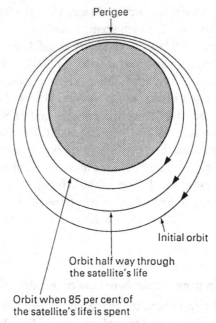

Perigee

Initial orbit

Orbit half way through
the satellite's life

Orbit when 85 per cent of
the satellite's life is spent

Fig. 2.6. Contraction of an elliptic satellite orbit under the action of air drag. The decrease in perigee height has been exaggerated for clarity.

decreases by a factor of 2.718, the exponential constant. Thus if ρ_p denotes the density at the perigee distance r_p, the density ρ at distance r is given mathematically by

$$\rho = \rho_p \exp\{-(r-r_p)/H\}.$$

But what was the value of H? No one knew for certain at that time, though we suspected it might be near 50 km for Sputnik 2. Taking H as constant was just a convenient simplification: probably it varied with height.

Assuming also that the density remained the same from day to day, we were able to express the satellite's lifetime L in terms of the initial (or current) orbital period T and its rate of change $\dot{T}$. The formula for L was only slightly dependent on H, if the eccentricity e was between 0.04 and 0.1, as for Sputniks 1 and 2. We found that

$$L \simeq -\frac{3eT}{4\dot{T}}\left(1+\frac{7e}{6}+\frac{H}{2ae}\right),$$

on ignoring small terms. For Sputnik 2, e was near 0.1 initially, a was 7200 km and, if with hindsight we take H as 50 km, the value of L from the equation is $-0.086\ T/\dot{T}$. Since T was 103.75 minutes initially and $\dot{T}$ was -0.05 minutes per day, the predicted lifetime was 180 days: this was 10% greater than the actual lifetime, but well within the margin of error to be expected. The main reason for the error was the assumption of a spherical atmosphere. It seemed more likely (as was later confirmed) that the density would depend on height above the oblate Earth, as it does near sea level. The perigee of Sputnik 2 moved from latitude 50° N initially to near the equator at the end of the life, and this would be expected to reduce the lifetime. In retrospect, this reduction is known to have been about 15%, and it was partly compensated by a general decrease in upper-atmosphere density as solar activity declined slightly in the early months of 1958.

The equation for lifetime was, and still is, useful at any time in the life. As the end of the life approaches, the relative error remains the same, so that the absolute error decreases. In an article on the progress of Sputnik 2 published in *Nature*[4] on 15 March 1958, my estimate for the decay date was 15 April, which was close to the actual date, 14 April. However, this was partly due to luck, because the air density fluctuates unpredictably from day to day by about 10%, thus usually limiting the accuracy of lifetime estimates to about 10% of the remaining life.

The theory also indicated how the decay rate $\dot{T}$ should vary with time t during the satellite's life. As a first approximation, $\dot{T}$ is proportional to $(1-t/L)^{-\frac{1}{2}}$, though in practice several smaller terms usually need to be added. The main results of the theory were given in an article published in *Nature*[5] in June 1958, and we were able to compare the values of decay rate predicted by theory with those determined by Doreen Gilmore from analysis of more than 400 observations used for the prediction service. The results are shown in Fig. 2.7, the errors in most of the observational values being about the size of the crosses. You can see large and systematic variations in drag. We had to abandon any idea that the atmosphere up there might be as placid as the proverbial mill-pond, and recognize instead that it was an arena of dynamic conflict. The variations were emphasized in the article in *Nature* on 15 March 1958, where I cautiously said that 'their interpretation should provide much scope for ingenuity'. In the paper in June we pointed out that 'an oscillation of period about 25 days is discernible'. The reason for this was to emerge later. We concluded in June that the variations were not likely to be due to changes in the effective cross-sectional area of Sputnik 2, because the observed brightness fluctuated with a period between $1\frac{1}{2}$ and 2 minutes. This suggested that the satellite was rotating many times during each circuit round the Earth: any changes in cross-sectional areas should therefore tend to average out.

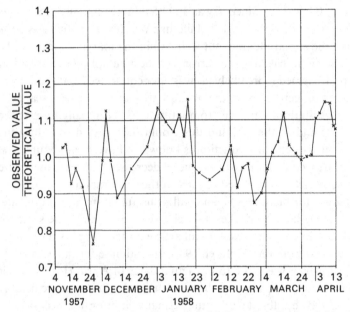

Fig. 2.7. Rate of decrease of orbital period for Sputnik 2. Ratio of observed values to theoretical (constant density being assumed for the latter). Revised version of diagram published in *Nature*, **181**, 1761–3 (1958).

Other fruits of the theory published in June 1958 included the useful result that, if the density stays the same from day to day, then e^2 decreases almost linearly with time, becoming virtually zero at the end of the life. Thus if e_0 is the initial value of e, its value at any later time t is given approximately by

$$e \simeq e_0 (1 - t/L)^{\frac{1}{2}}.$$

This was also a useful guide to the lifetime: for example, when e falls to half its initial value, 75% of the life has passed.

So, in the realm of atmospheric density, the main role of Sputnik 2 was to trigger the creation of new theory and then to expose the strong variations in density over timescales of days and weeks. The sources of these variations were to become apparent later in 1958.

The shape of the Earth revealed by Sputnik 2

The orbit of every satellite is sensitive to any irregularities in the Earth's gravitational pull, and it was widely expected that analysis of their orbits would reveal more about the gravity field, and hence the shape of the

Earth. This expectation became a reality with Sputnik 2, because the many accurate observations made by the RAE kinetheodolites were used by Robin Merson to determine accurate orbits.

In 1957 the only securely-established general feature of the Earth's shape was its oblateness or flattening. The flattening f is defined as the difference between the equatorial and polar diameters, divided by the equatorial diameter: the best value available for $1/f$ in the early 1950s was 297.1 ± 0.4, obtained by Sir Harold Jeffreys in 1948. This implied that the polar diameter was 42.94 km less than the equatorial diameter, and was the culmination of a series of estimates or measurements going back to the time of Sir Isaac Newton. The value favoured by Newton himself and given in the *Principia* was $1/f = 230$; the Cassinis in the early eighteenth century came up with a negative value, -95, implying that the Earth was like a lemon; expeditions to Peru and Lapland in the 1740s led to values between 179 and 266; George Everest in 1830 found $1/f$ to be 301, followed by 295 from A. R. Clarke in 1866, and 297.0 from J. F. Hayford in 1909.

It was a simple matter to calculate the expected value for the rate of rotation of the orbital plane for Sputnik 2 throughout its life from the formula for $\dot{\Omega}$ given on p. 18, and to compare this 'theoretical value' with the observed change as given by Merson's orbits, which were completed by May 1958. The result of the comparison was clear-cut: the observed value was consistently 0.7% less than the theoretical. This finding was announced at a meeting at the Royal Society in April, and numerical values were given in the published version of the paper on the effects of oblateness, which was revised on 15 May.

During the summer the values were improved, and the final result, that the ratio of theoretical to observed value was constant throughout the life at 1.0068, with an error estimated as only 0.0003, was published in *Nature*[6] on 6 September 1958. In this paper we said that if the difference was due to a wrong value of the second harmonic, the value of $1/f$ would need to be changed 'from 297.1 to 298.1 ± 0.1' (the value now accepted is 298.257). We also stated that the difference could possibly be caused by a wrong value of the fourth harmonic, for which a conventional value was assumed. To be sure of the culprit, results from another satellite at a different inclination were needed.

We also suggested that in future the Earth's gravitational potential U should be written in a form that took account of all the harmonics that might contribute, not only the second harmonic which gave rise to a flattening, but also the third harmonic, which was equivalent to a 'pear shape' or triangular tendency, the fourth harmonic, which was equivalent to a 'square shape', and so on. Of course we did not use these colloquial terms, but gave a mathematical formulation as an infinite series in the form†

$$U = \frac{\mu}{r}\left\{1 - \sum_{n=2}^{\infty} J_n \left(\frac{R}{r}\right)^n P_n(\sin\phi)\right\}.$$

In this equation, which is worth quoting because the J_n notation became standard, μ is the gravitational constant for the Earth, $398\,600$ km^3/s^2, and $P_n(\sin\phi)$ is the 'nth harmonic' or, to be more mathematical, the Legendre polynomial of degree n and argument $\sin\phi$, where ϕ is the latitude. The J_n are constant coefficients that might be evaluated up to higher and higher values of n as data from more and more satellites came in. Previously, only J_2 had been known. The even J_n, that is J_2, J_4, J_6, ..., go with harmonics having shapes that are symmetrical about the equator, and the odd J_n, namely J_3, J_5, J_7, ..., are associated with shapes that are not symmetrical north and south of the equator – for example, the fifth harmonic is like a flower with five petals. By combining all these harmonics we could hope for a much better representation of the real shape of the Earth. We chose the J_n notation in honour of Sir Harold Jeffreys. The results from Sputnik 2 indicated a value of J_2 near 1084×10^{-6}, as opposed to the previously accepted value of 1091×10^{-6}, if the conventional value of J_4 (namely -2.4×10^{-6}) was adopted.

Although I have implied that all this was plain sailing, that is not really so. Geodesists had been studying the shape of the Earth for more than 200 years, and they were not pleased to be told by young upstarts with no knowledge of geodesy that their accurate and laborious measurements were wrong. It was especially galling that the upstarts came from an aircraft establishment and had no idea how to measure distances accurately on the Earth. In the spring of 1958 we received a visit from the Geodesy Subcommittee of the British National Committee on Geodesy and Geophysics. The Chairman was Dr J. de Graaff-Hunter, FRS, and the distinguished members included Sir Harold Jeffreys, author of *The Earth*, and Brigadier Guy Bomford, author of *Geodesy*. We presented our results, but most of the geodesists could not believe that their standard value of J_2 was wrong and suggested that neglected higher harmonics might be causing error. This was a possibility that could not be completely excluded, but we thought it unlikely. Fortunately, Sir Harold Jeffreys welcomed our results: even he was sceptical, however, of the prospects of evaluating more than two or three of the J coefficients. Despite the scepticism of the real

† Our original equation for U used fM for μ and co-latitude instead of latitude.

professionals, we went ahead with the article for *Nature*. Fortunately the refereeing system had not then reached its present level of oppressiveness, and the paper was published without refereeing on the decision of the co-editor of *Nature*, Mr A. J. V. Gale, our great benefactor in those turbulent years.

Upper-atmosphere winds from Sputnik 2

People had expected that satellites might reveal more about the Earth's gravity field; but no one had predicted that it might be possible to measure how fast the upper atmosphere is rotating, and thus find the average west-to-east wind experienced by the satellite. Merson's orbits showed that the orbital inclination of Sputnik 2 decreased from about 65.33° initially to 65.19° at the end of the life. In the *Nature* article of September 1958 we noted this decrease and suggested that it might be caused by the sideways forces created by the rotation of the atmosphere. The subject was quantified in a theoretical study by Rex Plimmer, who derived the change Δi in inclination i produced by an atmosphere rotating at the same rate as the Earth, in terms of the change ΔT in orbital period. Though the full equation is too long to give, an approximate version is

$$\Delta i \simeq \tfrac{1}{3}\Delta T \sin i \cos^2 \omega,$$

where ΔT is in days and Δi in radians. So the observed decrease of inclination with time could be compared with the theoretical curve: the observed decrease proved to be about 25% greater than given by the theory. The slope of the theoretical curve 'would however be increased if there were a strong wind from west to east', we remarked in a note to *Nature*.[7] Numerically, it seemed that the upper atmosphere as experienced by Sputnik 2 was rotating about 25% faster than the Earth, thus implying west-to-east winds of about 100 m/s. This was the beginning of the use of orbits to determine upper-atmosphere winds.

The end of Sputnik 2

The final descent of Sputnik 2 through the atmosphere was just as spectacular as its launch. It could have come down in daylight without being seen, but instead the descent was in the evening. The night could have been cloudy, but was clear along the descent path. The location could have been remote, but the actual descent path ran from New York across the Caribbean Sea to near the mouth of the Amazon. The satellite was already self-luminous near New York, and it put on a fine display of fireworks as it crossed the Caribbean, where it was seen from many ships. Thanks to a call to shipping from Donald Sadler, we received at the RAE quite accurate observations from sixteen ships and a few stations on land, and we were able to determine the track and the height fairly well, Fig. 2.8. The appearance of Sputnik 2 became more bizarre as it proceeded south. At latitude 30° N the satellite was glowing a dull orange. At latitudes south of 20° N it had a long tail alive with sparks 'like those thrown off from a grinding wheel'. The decaying satellite took on almost every colour of the spectrum, grew a tail up to 100 km long and (when high in the sky) appeared brighter than Venus. Our analysis of the descent was published in *Nature* in August,[8] and my co-author was Doreen Walker, who had changed her name from Gilmore by getting married.

We never knew at the time how lucky we were with Sputnik 2. It was bright and easily observed. The air drag it suffered was severe enough to allow accurate measurement of the decay rate and its variations, yet the apogee was high enough to ensure a reasonably long life, allowing us to measure the effects of the Earth's gravitational field. The change in inclination due to atmospheric rotation was readily measurable, because of the large change in orbital period and the high inclination (65°). To this day, Sputnik 2 is probably the best satellite for orbit analysis that has yet been launched. As a final bonus, the orientation of the orbit on the days before decay was ideal for visual observing, and on the fine evening of 13 April many people in Britain went out to watch the satellite cross the sky, looking brighter than any star and only four revolutions before its fiery descent. In the subsequent thirty years I made more than 12000 satellite observations, but only once did I see a satellite nearer to decay than Sputnik 2.

By another lucky chance, Sputnik 2 came down a few hours before I was due to speak about the Russian satellites at what now seems an amazing conference entitled 'Britain enters the Space Age', at the Royal Festival Hall in London. The event was organized by the Air League of the British Empire – yes, Great Britain had an Empire then, as a few readers may remember – and the Festival Hall was filled with an enthusiastic audience, chiefly of schoolchildren. The Duke of Edinburgh spoke first; I was one of four subsequent speakers, Arthur C. Clarke being another. After the talks there was a question-and-answer session, with the speakers and Professor Massey answering the questions.

For me the Festival Hall lecture was memorable in several ways. First, the enthusiasm was something I never experienced again in Britain: in later years Britain seemed to be trying to escape from the Space Age, not enter

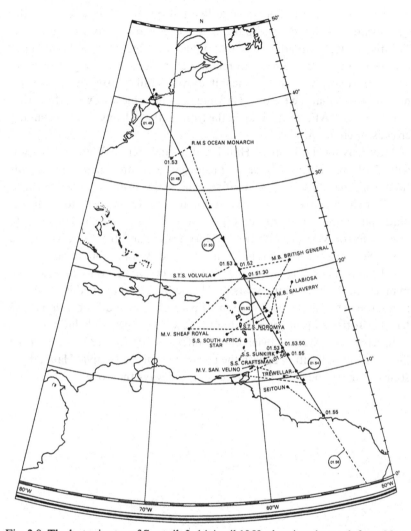

Fig. 2.8. The last minutes of Sputnik 2, 14 April 1958, showing the track from New York to South America. Published in *Nature*, **182**, 426–7 (1958). The ringed numbers give the best estimate of time in hours and minutes GMT. Unringed numbers are times given by observers, in hours, minutes and seconds GMT. The satellite's height is taken as 60 nautical miles at latitude 30° N, decreasing linearly with latitude to 40 nautical miles at latitude 10° N. ● indicates observer's position, ▲ indicates satellite's reported position.

it. Subsequent space scientists have never had thousands of cheering children to keep their spirits up. Secondly, the event made me realize the importance of luck and timing. By telling the multitude that Sputnik 2 had come down 10 hours ago in a brilliant display of fireworks over the Caribbean, I was adding a spurious touch of showbiz to my sober talk. Thirdly, the lecture made me see that public speaking is easy if the audience is large enough. People have a prejudice that the talk is going to be good because so many have come to hear it (as with a full theatre). What they did not know was that the Festival Hall had no slide projector: we had to bring one from the RAE, setting it up at the front of the balcony amid the milling crowds of children.

After leaving the Festival Hall I had a sandwich at Waterloo Station with Professor Harrie Massey. Thus began a secure yet usually distant friendship that lasted for twenty-five years and was of immense benefit to me. Our short conversation at Waterloo focused on the need for a British National Space Centre, preferably at the Royal Greenwich Observatory, and Professor Massey, with David Martin (the Executive Secretary of the Royal Society), took up this idea with Richard Woolley, but to no avail. It was 1985 before the idea was put into practice, though only in a low-key manner. In the intervening years Sir Harrie Massey (knighted in 1960) did in a sense act as a personal National Space Centre, and his death in 1983 left the British ship of space without a generally accepted helmsman.

In retrospect, that sandwich in the buffet at Waterloo, 12 hours after the fiery descent of Sputnik 2, with the razzmatazz of the Festival Hall fading, seems to mark the end of a chapter in more ways than one.

3

Full speed ahead, 1958–1960

Bring me my Spear! O clouds unfold!
Bring me my chariot of fire!

William Blake, *Milton*

The clouds had obediently unfolded to reveal that 'chariot of fire' over the Caribbean on 14 April 1958; but the descent of Sputnik 2 left us without any satellites to predict. The first US satellite, the pencil-shaped Explorer 1, had been launched on 1 February; the grapefruit-like Vanguard 1 followed on 17 March; and Explorer 3 on 26 March. But these three satellites were small and faint, and, with orbits inclined at less than 35° to the equator, they were far to the south and nearly always below the horizon for observers in Britain.

During this welcome respite there was time, on 22 April, for a visit to Herstmonceux, where the moated castle was worlds apart from the hotchpotch of rather ugly buildings at the RAE. (The contrast always startled me, even in later years.) Thus began a secure and friendly cooperation with the Royal Greenwich Observatory that flourished for more than thirty years, with benefit to both sides. The road back from historic Herstmonceux ran through Piltdown, a name redolent of even older times – or so it was thought until the Piltdown Man was exposed as bogus.

The hiatus in prediction did not last long, for Sputnik 3 was launched on 15 May, which was presciently marked in 1958 diaries as Ascension Day. We heard about the launch just before noon, and early that afternoon sent out the first set of predictions, which proved accurate to half a minute.

Sputnik 3 was a large satellite in the form of a cone about 2 m long and 1 m in diameter, designed to make scientific measurements of the space environment. As Fig. 3.1 shows, it was crammed with instruments that

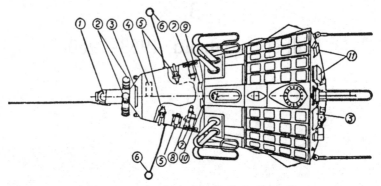

Fig. 3.1. Layout and instrumentation of Sputnik 3. Key: (1) magnetometer, to measure the intensity and direction of the Earth's magnetic field. (2) Photo-multipliers for measuring the sun's radiation. (3) Solar batteries. (4) Cosmic-ray meters. (5) Ionization manometers. (6) Ion traps. (7) Electrostatic fluxmeter. (8) Mass spectrometer. (9) Instruments for recording heavy nuclei in cosmic rays. (10) Apparatus for measuring intensity of primary cosmic rays. (11) Instruments to register the impacts of micrometeorites.

made up 968 kg of its total mass of 1327 kg. Sputnik 3 was accompanied in orbit by a separated rocket, which seemed to be (and was) similar to the rocket of Sputnik 1. The rocket was much brighter than Sputnik 3 itself, but was more difficult to predict because it suffered more severe air drag. So the rocket occupied most of our attention and analysis until its decay on 3 December 1958.

The prediction service

Our methods of prediction were based on those devised by the RGO; we merely made a few refinements when we took over. The weekly predictions for Sputnik 3 rocket (and previously for Sputnik 2) were of course graphical rather than computerized, and consisted of three items:

(1) A map of part of the Northern Hemisphere (issued only once, unless the recipient lost the original).

(2) A transparent track diagram, issued once a fortnight, to fit over the map and fixed to it by a drawing-pin through the north pole. The track diagram gave the position of the satellite at 1-minute intervals from apex, and its height at each of these points. The elevation of the satellite was indicated by broken lines.

(3) Setting data, a printed sheet listing times and longitudes at apex (point of maximum latitude north) for each day for every transit near Britain.

The procedure was to rotate the track diagram about the pin so that the thick 'zero-line' of longitude on the diagram was at the predicted apex

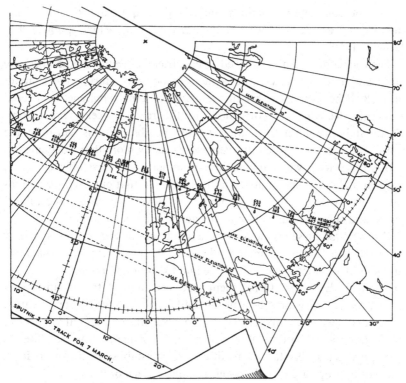

Fig. 3.2. Example of track diagram in use, set for revolution 1829 of Sputnik 2 on 9 March 1958. The transparent track diagram (for 7 March) is set with the apex longitude at 38° west, as predicted for revolution 1829. Observations from four stations in the UK are indicated, mostly at elevations between 20° and 40°.

longitude on the underlying map. This defined the satellite's track, and the maximum elevation was roughly indicated by the broken lines, or could be more accurately assessed by measuring the distance on the map.

Fig. 3.2 shows an example of the track diagram in use, for revolution 1829 of Sputnik 2 on 9 March 1958, apex time being 04 h 42 m UT.

To predict the apex times, we had to keep an accurate and up-to-date record of the satellite's orbital period, relying mainly on observations from the visual observers, who phoned in or sent the data by post. Our graph of orbital period was extrapolated graphically for a week or so ahead, and these extrapolated values were used to make the predictions of apex times. The future orbital heights were estimated from the semi major axis (obtained from the orbital period) and the perigee height, as given by the theory we were developing. The rotations of the orbital plane and of the perigee were calculated from the theoretical equations already given on

p. 18. The complete predictions could be done quite quickly on a desk-calculator. The methods were fully described in an article[1] published in January 1959.

The rocket of Sputnik 3 proved to be a tricky object to predict, because large and unexpected variations in drag occurred during the summer of 1958. Whenever the error in the predicted time threatened to exceed 1 minute, we sent out corrections to the regular weekly predictions. Fortunately, we never 'lost' the satellite: the observations sent in by the visual and kinetheodolite observers were usually more than adequate.

Sputnik 3 itself was much easier to predict. It was less affected by air drag, and also was transmitting radio signals, so that optical observations were not so vital. In June we estimated that it would remain in orbit for $1\frac{3}{4}$ years, and this was quite close to the actual lifetime of 692 days.

During 1958 these predictions were sent to about 180 addresses, mainly in the British Isles and Finland: about 3000 transparent track diagrams and 400 000 predictions of apex time and longitude were despatched. The excellent RAE Printing Department produced all the data in double-quick time, and we had willing help in the mailing of the sheets. The operation proceeded very smoothly. However, the work was substantial in terms of staff, paper and printing. With the prospect of many more satellites to come, it was decided that a scientific research establishment like the RAE was not appropriate for a long-term public service of this kind.

The Radio Research Station (RRS) at Datchet, near Slough, designated in 1958 as a World Data Centre for satellites, was chosen as the future home for the prediction service. The transfer was gradual. We had begun making predictions for the US satellite Explorer 4 launched on 26 July. Its orbit was inclined at 50.3° to the equator, and the satellite was therefore visible from Britain, though faint. This was the first satellite to be handed over to the RRS, in September 1958. At the end of October we handed over Sputnik 3, together with any new satellites that might be launched. We retained the rocket of Sputnik 3, now increasingly difficult to predict, until the end of its life on 3 December 1958.

Though the prediction service had left the RAE, it was only 24 miles away, and we maintained friendly links with the prediction staff at Datchet throughout their twenty-one years of operation.

While responsible for the prediction service, we received many requests for a table of all the satellites launched, with their sizes, shapes, orbits and likely lifetimes. In response Doreen Walker produced in July 1958 the first issue of what become known as the RAE Table of satellites: it was just one page, with all the archaic units of feet, pounds and nautical miles. We continued to compile the Table at the RAE, and by the end of 1960 it had grown to nine pages. It was still much sought after, as well as being useful in orbital researches.

The visual observations by volunteer observers were an unforeseen bonus in 1958. But it was not obvious whether they would be just a nine-months' wonder and would decline in numbers as the enthusiasm faded. Nor was it clear whether such visual observations would be numerous and accurate enough to determine orbits with good accuracy. High standards had been set in 1958–59 with the aid of the accurate observations from the kinetheodolites.

The number of visual and kinetheodolite observations in 1957–58 was 1281, of which 540 were of Sputnik 2, 525 of Sputnik 3 rocket, 156 of Sputnik 1 rocket and 60 of the Sputnik 3 spacecraft. These are the numbers of observations judged to be accurate to better than 1 s in time and 1° in direction. In 1959 the number of optical observations received by the RRS fell to 475, of which 415 were of Sputnik 3. This decrease was, however, largely due to the absence of easily observed satellites in 1959, and the total increased to 2153 in 1960. So by the end of 1960 it seemed that the volunteer observers would be able to provide observations in good numbers: whether the observations would be accurate enough for research purposes remained in doubt.

Variations of air density with time, and their origin

The erratic variations in the decay rate of Sputnik 3 rocket, which caused such trouble in prediction, had a deeper interest because their form might reveal the source of the irregularities. We were keeping a careful record of the decay rate and, once the lifetime was known (decay was on 3 December 1958), the observed decay rate could be divided by the value expected from theory, to indicate the variations in air density.

When these results were added to those for Sputnik 2, with a linking section for the US satellite Vanguard 1 (at a much greater height, 650 km), the picture that emerged was that shown as Fig. 3.3. (In fact these three diagrams were not printed in combined form until mid 1959, but the information was available in December 1958.) Fig 3.3. shows a pronounced periodicity in air density, with a period of about 28 days, and I concluded that these variations were likely to be attributable to solar disturbances, because the Sun rotates relative to the Earth once every 27.3 days, giving rise to well-known 27- or 28-day recurrence tendencies in sunspot numbers. This conclusion was first publicized in a note that appeared in the

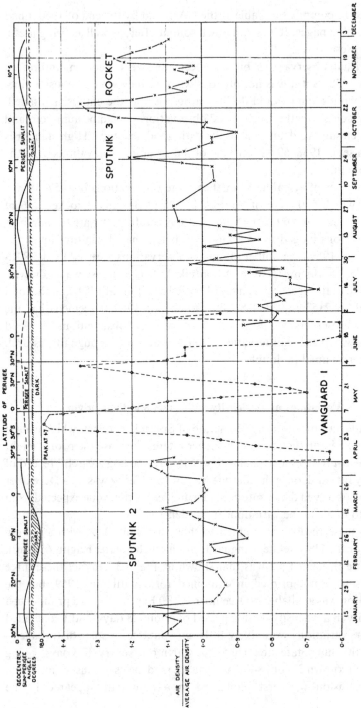

Fig. 3.3. Variation of air density during 1958 as given by Sputnik 2 and Sputnik 3 rocket (250 km height) and Vanguard 1 (650 km). There is a recurrence tendency with a period of about 28 days (the interval between the grid lines), but no sign of dependence on latitude, or of day–night effects.

newspaper *The Observer* for 30 November 1958. In January 1959 I went to the Royal Society to meet Robert Jastrow, a leading US space and atmospheric scientist, and told him of our conclusion. He was very sceptical; but within a month the hypothesis of solar control had emerged independently in the USA as a consequence of the elegant orbital analyses by Luigi Jacchia of the Smithsonian Astrophysical Observatory at Cambridge, Massachusetts.

Our results were published in *Nature* on 21 February 1959, together with a paper by Jacchia pointing to the same conclusion. We remarked that

The rather sharp onset of the increases in drag... suggests that the changes in air density are associated with the streams of particles which are projected radially from the Sun and sweep across the Earth at intervals of 27–28 days, giving rise to well-known periodicities in geomagnetic activity, cosmic rays and the aurora.[2]

Jacchia mentioned three difficulties with the solar radiation hypothesis but concluded:

the solar-radiation hypothesis still appears to me, in default of any better as the only plausible solution in sight.[3]

In a note added in proof on 14 January 1959 he referred to a comparison with solar 20 cm radiation made by W. Priester of Bonn; acting on this hint, Jacchia compared the densities with the solar 10.7 cm radiation, finding agreement 'little short of perfect'. It is curious that the hypothesis of solar control was previously suggested in June 1958 by Terence Nonweiler (*Nature*, **182**, 468). But his analysis was inconclusive, to the mystification of everyone, then and now.

The idea of solar control of the upper-atmosphere density was soon generally accepted, and many detailed comparisons were made, particularly by Jacchia, between the various indices of solar activity and the upper-atmosphere density as revealed by analysis of satellite orbits. The most useful of the solar indices proved to be the record of solar radiation at a wavelength of 10.7 cm, as measured at Ottawa by the National Research Council of Canada. This particular radiation, which comes through to the ground and is not absorbed by the atmosphere, has no direct influence on the upper atmosphere; but the 10.7 cm radiation is correlated with the solar ultra-violet radiation, which *is* absorbed by (and hence affects) the upper atmosphere. Ideally, the solar ultra-violet radiation would be accurately and continuously monitored on a routine basis from a spacecraft. But this has not yet happened, and so the 10.7 cm radiation has remained the favourite index over the time covered in this book.

Fig. 3.4 shows one of Jacchia's later comparisons,[4] in 1961, when the

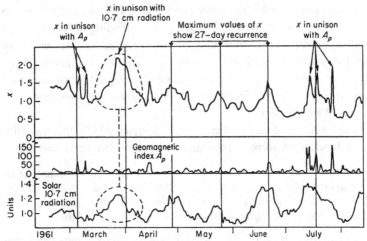

Fig. 3.4. Rate of decrease of orbital period due to air drag, *x*, for Explorer 9, which provides a measure of the air density at heights near 600 km. Sometimes *x* is in unison with the geomagnetic index *Ap*, sometimes with the 10.7 cm radiation. (Adapted from a paper by Jacchia,[4] which gives details of the units, etc.)

technique was well established. Most of the variations in air density are paralleled by variations in the 10.7 cm radiation, though there are some sudden increases in density that correlate better with geomagnetic storms, as indicated by the 'planetary geomagnetic index' *Ap* also plotted on the diagram. The index *Ap* is a measure of the disturbance in the Earth's magnetic field, averaged over a number of observatories round the world. The disturbances are caused by streams of particles from the Sun which sweep across the Earth and induce complex electrical currents in the outermost atmosphere (the magnetosphere). So diagrams such as Fig. 3.4, and its less-clearcut predecessors, showed that the solar control over the upper atmosphere has a dual personality, as it were. The ultra-violet radiation tends to create the longer-term variations, on time scales of a few days, a week or a month; but these variations are punctuated with sharper peaks correlated with geomagnetic storms. Attempts to model these complex effects mathematically began in 1960, and still continue today, with increasing but not complete success. One difficulty is that the effects depend much on chance, such as the position of the Earth in its orbit relative to the site of the solar eruption, and the direction and speed of the erupting plume.

The variation of air density with height

Sputnik 1 had given a value of air density at a particular height a little above the perigee (see p. 29). By making similar calculations for each new satellite launched, covering a variety of perigee heights, we were able to produce a tentative picture of how density varied with height.

Numerical values of air density could be found from the decay rate $\dot{T}$ of satellites by means of the theory we were developing. The theory led to a fundamental equation linking the air density at perigee ρ_p with $\dot{T}$, various orbital parameters and the area/mass parameter Δ. This equation is

$$\rho_p = -\frac{\dot{T}}{3\Delta}\left(\frac{2e}{\pi aH}\right)^{\frac{1}{2}}\left\{1-2e-\frac{H}{8ae}\right\}$$

and it applies for values of e between about 0.02 and 0.2, ignoring terms of order e^2 and H^2/a^2e^2. Good values of $\dot{T}$, a and e were usually available: the problems were the values of Δ and H.

The area/mass parameter Δ is given by SC_D/m, where S is the mean cross-sectional area of the satellite perpendicular to its direction of travel, m is its mass and C_D the drag coefficient. For all the satellites launched in 1958 (except Sputnik 3 rocket), the mass m was known, and so were the satellite's size and shape. If the satellite is spherical, S is easy to calculate, being just $\pi \times (\text{radius})^2$. But problems arise if, as usually happens, the satellite is non-spherical and rotating. Fortunately the rotation would not be completely random, for, as a result of what might be called the law of cosmic laziness, a rotating body not constrained by strong forces will tend to rotate as slowly as it can. A slender cylindrical satellite spins more slowly if it rotates like a propeller, rather than like a screwdriver, just as a competition skater spins more slowly with arms outstretched horizontally than with arms down close to the body. Most of the satellites of 1958 were roughly cylindrical, and their rotation can be envisaged by sticking a pin into a wooden pencil halfway along and perpendicular to its length, and then rotating the pin between your fingers. The pencil (satellite) may rotate like a propeller, the pin being along the direction of motion; it may tumble end-over-end, with the pin perpendicular to the direction of motion; or, more likely, the pin may point in neither of these directions but somewhere 'betwixt and between'. For a cylinder of length l and diameter d, with $l/d = 3$, it turns out that $S = 0.81\, ld$ if the satellite tumbles end-over-end; while $S = ld$ for propeller-like motion. In 1958 we took S as the mean of the values in these two extreme modes. Though this implied maximum errors of nearly 10%, the extremes were very unlikely to occur, and we usually quoted 5% as the standard deviation.

This gives a value for S/m, but what about the drag coefficient C_D? A value somewhere near 2 was thought to be likely in 1957, but little work on the subject had been done. In the summer of 1958 the RAE was fortunate to recruit Graham Cook, a young graduate with a background in aerodynamics, who would be able to work full-time on satellite orbits and aerodynamics. His first task was to survey the literature on the drag of objects in free-molecular airflow, and to recommend a value of C_D for use in our determinations of air density. After a thorough study, Graham came up with the value $C_D = 2.2$, with an estimated error of about 5%, to cover the majority of the satellites currently being analysed. This value has endured as the basic standard for thirty years, though subsequent work showed that different values would be appropriate for later satellites in orbits with higher perigees. In 1958 it was widely believed that the drag of satellites included an important contribution from 'electrical drag': but we ignored this possibility and just took the aerodynamic C_D of 2.2.

With the value of Δ decided on, there remained a second difficulty over the equation for ρ_p: the value of the scale height H was not known at all accurately in 1958. We soon found a way round this problem – by evaluating air density at a height $\frac{1}{2}H$ *above* perigee. You choose the best value of H you can, and call it H^*. Then, if H^* is not in error by a factor of more than 1.5, the density ρ_A at a height $\frac{1}{2}H^*$ above perigee is given by

$$\rho_A = -\frac{0.158\dot{T}}{\Delta}\left(\frac{e}{aH^*}\right)^{\frac{1}{2}}\left\{1-2e-\frac{H^*}{8ae}\right\},$$

with error less than 5% if e is between 0.025 and 0.15, as for most of the satellites of 1958. This equation, published[5] in May 1959, may seem so convenient as to be suspicious. The reason for its success is that the satellite samples air density at all heights between the perigee and the apogee, though heights near the perigee are obviously most important. Thus, in effect, there is an 'average height' at which the density is being evaluated, which is likely to be not far above the perigee. Fortunately it happens that, for a usefully wide range of values of eccentricity, this 'average height' is half a scale height above the perigee. The equation for ρ_A had long and useful service in determining upper-atmosphere density, after a slight alteration to allow for the effect of atmospheric rotation (Δ has to be replaced by $\delta = F\Delta$, where F is a factor that is nearly always between 0.9 and 1.0).

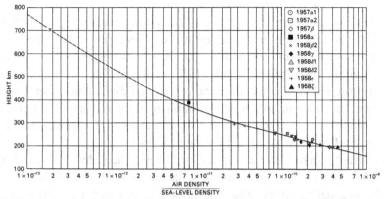

Fig. 3.5. Values of air density obtained[5] from satellites launched in 1957 and 1958. To convert the scale to absolute density, note that air density at sea level is 1.2 kg/m³.

The equation was applied early in 1959 to all the satellites launched before then, and the resulting curve of density versus height is shown as Fig. 3.5.† This picture of density versus height was significant for several reasons. First, it offered the best indication of the variation of density with height in 1958. Secondly, there were no serious anomalies, thus suggesting that the assumptions made in calculating cross-sectional area, etc., were probably correct. Thirdly, the slope of the curve gave an indication of the hitherto-unknown values of the scale height H. It will be remembered that H is the height increase needed to reduce the density by a factor of 2.7, and Fig. 3.5 shows that the relative density is 2.7×10^{-11} at 300 km height and 10^{-11} at about 360 km height. So the scale height here is about 60 km. As the curve steepens at greater height, we presumed that H increased with height. If the curve drawn is taken as correct, H increases from about 40 km at a height of 200 km to about 100 km at a height of 700 km. Though these values were not likely to be accurate, they indicated that our theory for the effect of the atmosphere on orbits needed to be modified, because we had so far assumed that H was constant.

Real life breaks in

In September 1959 the Tenth International Astronautical Congress was held in London at Church House, Westminster. This Congress was – and

† In Fig. 3.5 the satellites are specified by their international designations rather than their popular names. Thus 1957β is Sputnik 2, 1958α is Explorer 1, and so on. In this book the popular names are used whenever possible; the appropriate international designation is given in the index.

still is – the most important annual gathering devoted to space exploration in general, and was an ideal forum for presenting our results, which had not yet received any international 'exposure'. I spoke about the variations of air density with time and with height, the results on atmospheric winds from Sputnik 2 and Sputnik 3 rocket (which both indicated strong west-to-east winds), and the studies of the gravity field to be mentioned later in this chapter. This presentation[6] was favourably received and helped to put the RAE 'on the map' internationally. A week later I gave a somewhat similar talk at the British Association meeting in York.

In the autumn of 1959 Sir George Gardner was succeeded as Director of the RAE by Professor James Lighthill, one of the leading mathematicians of the twentieth century, who had been elected a Fellow of the Royal Society in 1953 at the age of twenty-nine. This was another lucky chance for me, because Lighthill, as a mathematician, was likely to be interested in our work. Also he had been my supervisor in 1946 when I was an undergraduate at Trinity College, Cambridge.

In February 1960 Professor Alla Massevitch came from the USSR to lecture in Britain on the scientific results of space exploration. She was Vice-President of the Astronomical Council of the USSR Academy of Sciences, and was much concerned with satellite tracking and orbits. On 24 February, at Lighthill's invitation, she gave a lecture at the RAE to an audience of several hundred of the staff, and in perfect English. I had wondered whether we should be blamed for making piratic use of the satellites so expensively launched by the USSR. But she was most appreciative of our work, and very kind and generous in my subsequent discussion with her. Though I didn't realize it at the time, her visit also helped to maintain support for our work: already some people were asking why the Royal *Aircraft* Establishment was meddling in such way-out subjects as space.

Another administratively-helpful event had occurred in the summer of 1959 when, to my great surprise, I was awarded the Bronze Medal of the Royal Aeronautical Society. I felt rather an imposter, because our work was not really aeronautics; but this recognition by the aeronautical 'establishment' helped to make us more respectable among the dominant 'aircraft side' of the RAE. Not just wild space-people after all. The presentation of the Medal was the first and only time I have ever worn 'evening dress' – borrowed from Clifford Cornford a few hours before. I resolved not to wear a penguin suit again, and never have done, thereby no doubt missing some 'honours and awards' that depend on conformist dressing. But just think how many deadly dull dinners I have avoided!

The day-to-night variation in air density

Now for a different kind of black and white, the contrast between day and night. As time passes, the perigee point of a satellite moves slowly in local time from day to night and back again. For Sputniks 1–3, at inclination 65°, the perigee moved at a rate of a little less than 5° per day on this day-to-night trek, completing a cycle in about 80 days. If the density changed greatly between day and night, there would be great changes in the decay rate. No such changes were observed, though there might have been small ones overwhelmed by the irregular variations with time already discussed. So it seemed that, at heights near 200 km in 1958, day-to-night changes in density were quite small.

Would the picture be different at greater heights? The only 1958 satellite with a perigee above 400 km was Vanguard 1. For this satellite, at inclination 34°, the day-to-night cycle took more than a year to complete, and the perigee did not enter the Earth's shadow until April 1959. Within a few months the decay rate of Vanguard 1 decreased greatly: so it seemed that, at heights near 600 km, the air density was much greater by day than at night. Jacchia, who was working on the orbit of Vanguard 1, was the first to notice the effect. The perigee of Vanguard 2, launched in February 1959, entered the Earth's shadow in July 1959 and there was a similar decrease in decay rate. Several other examples were available by mid 1960, as Fig. 3.6 shows. All the satellites decay more slowly when the perigee is in darkness, the change being greatest for the Vanguard satellites, which have the highest perigees.

As a result of this discovery, the values of density determined at heights above 300 km had to be divided into two groups labelled 'midnight' and 'midday'. Fig. 3.7 shows the first profile of density of this kind, published in *Nature*[7] in June 1960 and based on results from twenty-one satellites. The obvious feature of this diagram is the indication that the maximum daytime density is much greater than the minimum night-time density at heights of 500–700 km. Also pointed out in the paper was another significant feature, not apparent at first sight, a slow decrease of density with time at the lower heights. The twenty-four points below 300 km can be divided into two groups, for October 1957 to January 1959, and for August 1959 to March 1960. The fourteen points in the first group are on average 10 % higher than the curve, and the ten points in the second group are 10 % lower than the curve. This probable 20 % decrease in density was expected, because solar activity declined considerably between 1958 and the end of 1959, and the linking of the weekly/monthly variations with

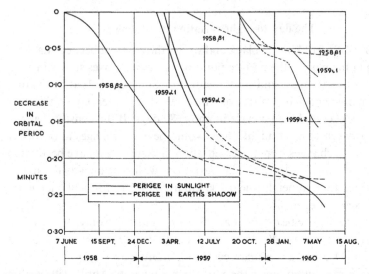

Fig. 3.6. Illustrating the effect of perigee entering shadow. The orbits decay more slowly when perigee is in darkness, the effect being greatest for Vanguard 1 (1958β2) and Vanguard 2 and its rocket (1959α1 and 2) which have the highest perigees (650 km).

solar activity seemed likely to apply also for the longer-term variations over a sunspot cycle.

Long-term changes in density dependent on solar activity

If the air density at heights of 200–300 km decreased by 20 % between 1958 and late 1959, much greater decreases were likely at greater heights. When data became available, future profiles of density versus height would have to be divided according to *both* the level of solar activity *and* the local time at the perigee (day or night).

New data were added, to produce profiles embodying this double division, in time for a paper presented at a Symposium on Aeronomy at Copenhagen in July 1960. The curve for midday in Fig. 3.7 was split into two, for early 1959 and mid 1960, and the curve for midnight was labelled '1959–60'. There would have been a fourth curve if any midnight values had been available for dates earlier than the middle of 1959. We commented as follows:

The maximum daytime and minimum night-time values of density given by the various satellites form a consistent pattern, with the maximum exceeding the minimum by a factor which rises from about 1.5 at 400 km to between 6 and 10 at 600 km. There is some indication that the daytime density has decreased, between

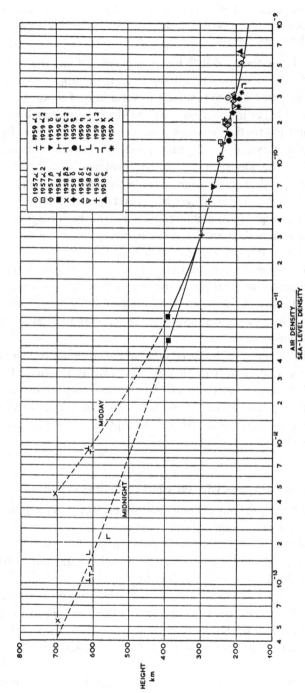

Fig. 3.7. Upper-atmosphere density obtained[7] from the orbits of twenty-one satellites launched in 1957–59, showing the day-to-night variation in density.

1958 and 1960, by perhaps 20–50 % at a height of 600 km. Such a decrease would be expected in view of the decrease in solar activity since 1958, and is in accord with findings at heights below 300 km.[8]

The 'songs of innocence' of 1958 had now developed into the 'songs of experience' of 1960. Experience had shown that the upper-atmosphere density at heights near 200 km is subject to variations of up to ± 25 % in response to (a) changes in short-wave solar radiation (presumably in the ultra-violet) and (b) particle streams from the Sun, as indicated by the geomagnetic planetary index. At greater heights the variations in density are stronger. In the longer term, the upper-atmosphere density was responding to the decline in solar activity between 1958 and 1960, decreasing by perhaps 20 % at 200 km height and by up to 50 % at 600 km. Also there is a strong day-to-night variation in density at heights above 300 km, the maximum daytime density being up to 10 times greater than the minimum night-time density at heights near 600 km.

Other variations were suspected by 1960, but most of these suspicions evaporated, and only one has stood the test of time. This is a variation dependent on the day of the year, exhibiting 6-monthly and 12-monthly oscillations, and usually known loosely as the 'semi-annual variation'. It will figure in Chapter 4.

In 1959 and 1960 many of the variabilities of the upper atmosphere were being uncovered and measured, though their origins remained largely mysterious. But we were very conscious that the theory so successfully applied in probing the upper atmosphere was much over-simplified: because of the complications of the real atmosphere, the theory urgently needed improving.

Creating new theory for the effect of air drag on orbits

As explained in the previous chapter, the orbital theory underlying this research on the upper atmosphere was based on the simple assumptions of a constant scale height H and a spherically symmetrical atmosphere. But we now knew that H was not constant: it increased considerably as height increased. We also knew that the atmosphere was, like the Earth, oblate rather than spherical; and the oblateness was modified by a day-to-night variation, which you can think of simply as a daytime bulge of density towards the Sun – as if the Sun were creating an attraction and pulling up the air into a hump (though that is not really the mechanism).

The scientific world was most appreciative of the new results on the properties of the upper atmosphere. We were applying new methods to new data. Invitations to write articles and attend conferences were embarrassingly numerous. The work applauded was quite easy – applying theoretical formulae to the available data with the help of a desk calculator (and six-figure tables of sines and cosines).

The really difficult and interesting work, which went unapplauded (and sometimes almost unnoticed, I suspect), was the creation of the orbital theory. David Leslie, my first collaborator in the theory, left the RAE before the end of 1958, and from then onwards it was Graham Cook and Doreen Walker who worked with me on the theory.

The results of the work could often be expressed in fairly compact equations such as those quoted in Chapter 2, but we ploughed through a heavy mass of algebra *en route*, with ever-present opportunities for mistakes. An exact theory was rarely possible, and the aim was to create an approximate theory that would be accurate enough for practical purposes and would carry within itself estimates of the magnitudes of neglected terms. We began with orbits having eccentricity e less than 0.2, which included all those launched before August 1959 (and most of the later ones). In developing the theory, we used expansions in powers of e and then neglected terms of order e^4, which should be less than 0.002, and higher powers of e. In the theory for an oblate atmosphere, we defined an oblateness parameter c that was also less than 0.2, and neglected terms of order c^4, c^3e, c^2e^2, and so on.

To help in working out the theory we divided the satellite's lifetime into two parts, which we called 'Phase 1' and 'Phase 2'. The changeover occurred when the quantity $z = ae/H$ decreased to the value 3. As H/a was usually near 0.008, the orbit usually entered Phase 2 when e was reduced to about 0.024. If the initial eccentricity was quite large, perhaps 0.1, Phase 1 would apply during most of the lifetime (94% if $e = 0.1$ initially). If the initial eccentricity was less than 0.024, however, the whole life would be under Phase 2. This division was, in a sense, a blemish in the beauty of the theory, but was unavoidable because terms in e/z, e/z^2, and so on, arose in the equations, and we generally needed to ignore terms in e/z^3 ($= H/az^2$), which would exceed 0.001 if z was much less than 3.

Our plan of campaign in developing the theory (Parts I and II) was for Graham Cook to 'take the plunge' and map out a tentative (or possibly definitive) course. He passed his work to me, usually a few pages at a time, and I went through it in detail, sometimes agreeing, sometimes modifying and occasionally suggesting a different approach. Our main arguments were over small terms: should they be kept explicitly or written as 'order terms'? It was difficult to know whether an 'order term' might be required explicitly at a later stage. As each section was agreed between us, a definitive version was written out by one of us, checked by the other, and then passed to Doreen Walker for independent checking. The opportunities for error were innumerable, and further mistakes were sometimes found at this stage in small terms: the subsequent algebra then had to be re-worked. As far as we are aware, no algebraic errors got through to the published papers for Parts I–III (though a small error in a subsidiary approximation was later found in Part IV). In the final papers the wording was mine, Doreen drew the diagrams, and the equations were everyone's responsibility.

Part I of the theory, for a spherically symmetrical atmosphere, appeared as an RAE Technical Note[9] in September 1959, and a shortened version was published in the *Proceedings of the Royal Society* in 1960. As the technique was the same in the seven subsequent parts, it may be worth describing briefly here, for the benefit of mathematical readers (others can skip the rest of this paragraph). We began from Lagrange's planetary equations, which give the rates of change of a and e due to drag. These were then expressed in terms of the eccentric anomaly E, and integrated round the orbit from $E = 0$ to $E = 2\pi$, to evaluate the changes in one revolution, Δa and Δx, in the semi major axis a and the quantity $x = ae$, in terms of Bessel functions $I_n(z)$ of the first kind and imaginary argument, of order n and argument $z = ae/H$. Dividing Δa by Δx gives a fairly simple equation of fundamental importance,

$$\frac{da}{dx} = \frac{I_0(z)}{I_1(z)} \cdot \frac{H}{a} - \frac{3H}{2az} + \frac{e^2}{z} + O\left(e^4, \frac{e}{z^3}\right),$$

in which the neglected terms are less than 0.002 if $e < 0.2$ and $z > 3$ (Phase 1). This equation carries within itself an accurate measure of the relative rates of decrease of apogee and perigee heights, and can be integrated to produce the equations already given in Chapter 2, and others. Of these, the most important specifies the decrease in the perigee distance r_p as the satellite's life progresses. With suffix 0 denoting initial values, the full equation for Phase 1 is

$$r_{p0} - r_p = \tfrac{1}{2}H\left\{\left(1 - \frac{5H}{2a_0}\right)\ln\frac{e_0}{e} - (e_0 - e)\left(1 - \frac{3H}{4a_0} \cdot \frac{1+e_0}{ee_0} - \frac{e+e_0}{2}\right) + O\left(e^3, \frac{1}{5z^2}\right)\right\}.$$

For a typical low-perigee orbit, with $H = 40$ km, $e = 0.1$ and $a = 7300$ km, the neglected terms are of order 20 m: so the equation gives the change in perigee distance due to air drag with excellent accuracy if H is known. Also, as the equation links the geometrical quantities r_p and e, and does not involve time, the result is unaffected by day-to-day changes in density, except in so far as these affect H (for which a mean value would be needed).

The long equation above, if reduced to its simplest form, indicates that the decrease in perigee distance is $\frac{1}{2}H\ln(e_0/e) + O(He)$. This gives a good idea of how the orbit behaves as its life progresses. As explained in the previous chapter, e^2 decreases linearly with time: so 75% of the life has passed when e decreases to half its original value e_0. At this time $\ln(e_0/e)$ is equal to $\ln 2$, which is 0.69, and the simplified equation shows that the decrease in perigee distance is about $0.35H$, or 14 km if $H = 40$ km. The decrease in apogee height is of course much greater: for example, if $a_0 = 7300$ km and $e_0 = 0.1$, the apogee height decreases by 784 km while the perigee height decreases by 14 km. This gives an accurate answer to the original question on how quickly the orbit becomes circularized.

The equation for the decrease in perigee distance has two other practical implications. Because the decrease is so slow, the satellite will sample the air density (at heights near the perigee) at almost the same height throughout the first 75% of its life. Thus, with a small correction for the change in height, one satellite can monitor the air density at a specific height throughout the greater part of its life. And, unlike the readings from instruments aboard satellites, the air density determined in this way is free from inaccuracies due to instrumental drift. The second implication of the equation for perigee distance is that, if the orbit is accurate, an accurate value can be found for the decrease in perigee distance, and hence an accurate value of the scale height H. This was to prove more difficult than expected, because of the need to remove the changes in perigee height caused by other forces, particularly the odd harmonics of the gravity field and the perturbations caused by the Sun and Moon.

The equations already given in this and the previous chapter specify the decrease of orbital period with eccentricity, and the decrease of eccentricity with time. Part I of the theory thus provided a complete solution, of an accuracy quite adequate in practice, if the atmosphere was spherically symmetrical and of constant scale height H. Eccentricities greater than 0.2 were not covered, however. Thus there were already three further tasks to be tackled tomorrow, or preferably yesterday: oblate atmosphere, high eccentricity and non-constant scale height. These were to be the topics of Parts II, III and IV of the theory.

Part I did have a more practical consequence: it provided the motive for my first journey outside England, to Nice for the first International Space

Science Symposium in January 1960. This conference was organized by the international Committee on Space Research, always known as COSPAR, which was set up in 1958 to foster international cooperation in space research: it has done so most successfully ever since in an admirable atmosphere of amity. International Space Science Symposia were organized annually by COSPAR from 1960 until 1980, and every two years thereafter. The papers presented were published in massive tomes called *Space Research* (1960), *Space Research II* (1961) and so on to *Space Research XX*, recording the 1979 conference. Since then, the papers have appeared in a 'periodical' entitled *Advances in Space Research*. At the Symposium in Nice, I spoke about Part I of the theory, and the written version was the second paper in the inaugural volume *Space Research*. At Nice I met for the first time many of the American, Russian and European scientists who were to be colleagues for the next twenty-five years. As well as bringing scientists together, COSPAR was organizing international satellite observing by arranging standard codes for reporting observations and exchanges of site coordinates. Such cooperation was essential in a subject so desperately dependent on good geographical coverage. It was ironic that world-wide activity was so important for the work of someone who had been stuck in one small country for all his 32 years. I was disappointed by Nice, probably through having conjured up too favourable an image of the unknown 'abroad': the raw sewage streaming out from the Promenade des Anglais was particularly disillusioning.

By the time of the Nice Symposium we were well into Part II of the theory, which took a more realistic atmospheric model, namely an oblate atmosphere: when a numerical value was needed, the ellipticity ε was taken the same as that of the Earth (0.00335). The theory proved to be very lengthy, and the extra terms to be added to the spherical-atmosphere results involved Fresnel integrals and cosine integrals, which are rather awkward to use. However, some simplifications were possible. The effects of atmospheric oblateness were expressed via the quantity c defined by $c = (\varepsilon/2H)r_{p0}\sin^2 i$, and the extra term to be added to the equation for da/dx on p. 62 is $(2c/z^2)\cos 2\omega$. The terms to be added to the long equation for $r_{p0} - r_p$ on the same page include Fresnel integrals if ω varies widely; but if ω is nearly constant, the extra term within the curly brackets is $(4c/z_0 z)(z - z_0)\cos 2\omega$. This second paper, Part II of the series, having the subtitle 'with oblate atmosphere', was issued as an RAE Technical Note[10] in December 1960 and published in the *Proceedings of the Royal Society* in 1961, in a much shortened version.

These descriptions of orbital theory will have been difficult to read and understand, but I can scarcely ignore work that occupied so much of our time. It was the key to the researches on air density, as I have already remarked; but it was also essential when evaluating the gravity field, because we needed to be able to remove the orbital effects of air drag accurately and reliably.

Gravity field and shape of the Earth

Our analysis of Sputnik 2 had shown that the second harmonic J_2 in the gravity field was slightly smaller than the value previously accepted, as described in Chapter 2. What we hoped to be able to do now was to evaluate more of the even-degree J-coefficients, that is, J_4, J_6, J_8, ..., and thus to determine the Earth's gravity field more accurately and in more detail. The observed rate of rotation of the orbital plane of any satellite gives a value for a 'lumped' even harmonic, of the form $J_2 + AJ_4 + BJ_6 + ...$. Here A, B, ... are numerical factors which are different for each different orbital inclination. These observed values of lumped harmonics, obtained from satellites at several different inclinations, give several linear equations between the J coefficients, from which a number of the coefficients can be calculated, if the higher ones are ignored.

The first such evaluation, presented at a Royal Society meeting on 12 November 1958 (published[11] in 1959) relied on results from Sputnik 2 at 65° inclination and Vanguard 1 at 34°. Values of J_2 and J_4 were determined, on the assumption that J_6, J_8, ... were zero, as $10^6 J_2 = 1083.3 \pm 0.7$ and $10^6 J_4 = -1.1 \pm 0.9$.† The resulting value for the flattening f was $1/f = 298.2 \pm 0.1$. Jacchia made a similar determination at about the same time, obtaining $10^6 J_2 = 1082.7$ and $10^6 J_4 = -2$. If his errors were similar to ours, the two pairs of values were consistent; but the effect of ignoring J_6 was questionable.

Next we added preliminary orbital data for Explorer 4 at 50° inclination, and derived $10^6 J_2 = 1083.1 \pm 0.2$, $10^6 J_4 = -1.4 \pm 0.2$, and $10^6 J_6 = 0.1 \pm 1.5$. These values were published (in terms of J and D) in *Nature*[12] on 28 March 1959, with the flattening given as $1/f = 298.20 \pm 0.03$.

These numerical values become tedious, and I shall only cite one more set, using Explorer 7 instead of Explorer 4. The crucial observational value here was the orbital inclination of Explorer 7, a faint satellite that I observed with binoculars from my back garden. As the inclination of Explorer 7 was near 50°, the observations were nearly overhead and

---

† The $\pm$ errors quoted here and subsequently are the estimated standard deviations, sometimes abbreviated to 's.d.'.

Table 3.1. *Values of J_2, J_4 and J_6 determined from satellite orbit analysis, with authors' error estimates, and recent values for comparison*

Author and year of presentation or publication	$10^6 J_2$	$10^6 J_4$	$10^6 J_6$	Published in:
Jeffreys, 1952	1091 ± 3	$[-2.4]$		*The Earth.* 3rd edn, CUP, p. 184
Merson & King-Hele, 1958	1084	$[-2.4]$		*Nature*, **182**, 640–641
Jacchia, 1958	1082.7	-2		Smithsonian Astrophys. Obs. Spec. Rpt 19
King-Hele, 1958	1083.3 ± 0.7	-1.1 ± 0.9		*Proc. Roy. Soc.*, A**253**, 529–538
King-Hele & Merson, 1959	1083.1 ± 0.2	-1.4 ± 0.2	0.1 ± 1.5	*Nature*, **183**, 881–882
O'Keefe *et al.*, 1959	1082.5 ± 0.1	-1.7 ± 0.1		*Astronom. J.*, **64**, 245–253
Buchar, 1960	1083.0 ± 0.4	-1.1 ± 0.5		Paper at IUGG Assembly 1960
King-Hele, 1960	1082.79 ± 0.15	-1.4 ± 0.2	0.9 ± 0.8	*Nature*, **187**, 490–491
Kozai, 1961	1082.19 ± 0.03	-2.1 ± 0.1		*Astronom. J.*, **66**, 8–10
Michielsen, 1961	1082.7	-1.7	0.7	*Advances in the Ast. Sci.*, Vol. 6
D. E. Smith, 1961	1083.15 ± 0.2	-1.4 ± 0.3	0.7 ± 0.6	*Planet. Space Sci.*, **8**, 43–48
GEM T3, 1992	1082.63	-1.62	0.54	NASA Tech. Memo 104555

proved accurate enough to give a good value for the inclination, namely $50.27° \pm 0.01°$. The values obtained for the J coefficients were $10^6 J_2 = 1082.79 \pm 0.15$, $10^6 J_4 = -1.4 \pm 0.2$, and $10^6 J_6 = 0.9 \pm 0.8$. These values were published in *Nature*[13] on 6 August 1960, with a value of $1/f$ given as 298.24 ± 0.02, which is quite close to the value accepted today, namely 298.257. The methods were described fully in a paper published in a special 'Harold Jeffreys issue' of the *Geophysical Journal* of the Royal Astronomical Society.[14]

By this time values had also been published by John O'Keefe and Emil Buchar: these, and three further evaluations, in 1961, by Yoshihide Kozai, Herman Michielsen and David Smith, are listed in Table 3.1, with the best values available today, from the Goddard Earth Model T3 (GEM T3) derived at the NASA Goddard Space Flight Center. The GEM T3 values have been rounded off and should be exact, to the number of figures given.

Nearly all these early determinations of even harmonics stand up well when compared with the currently accepted values. If you give an error (s.d.) of 0.1 to Michielsen's values of J_2 and J_4, and 0.2 to his J_6, you find that all the J_2 values (except Kozai's) are within 2.5 s.d. of the 1992 value, and all the J_4 and J_6 values (except Kozai's) are within 1 s.d. of the currently accepted values. All in all, the evaluation of J_2, J_4 and J_6 seems in retrospect a very successful example of scientific endeavour.

Unfortunately things did not work out quite so well at the time, when Kozai's values were widely adopted because of their apparently excellent accuracy, and found their way into several textbooks and computer models of orbits in the early 1960s. In retrospect, Kozai's value of J_2 is seen to be in error by 14 s.d.; or, to put it more positively, a standard deviation about 10 times larger would have been more realistic for J_2; and about 5 times larger for J_4. Here again, fortune smiled: most US scientists adopted Kozai's values; we stubbornly clung to our own, and thereby avoided errors in orbital models.

The harmonics of Table 3.1 measure variations of gravity with latitude, and are called *zonal* harmonics, because they measure the changes between the torrid, temperate and frigid zones of the Earth. Variations of gravity with longitude, which are much smaller than with latitude, are averaged out in zonal harmonics.

So far I have ignored the odd zonal harmonics, having coefficients J_3, J_5, J_7 and so on. The even zonal harmonics are symmetrical about the equator; the odd ones are asymmetrical. As Fig. 3.8 shows, the third harmonic corresponds to a rounded triangle in form, and is usually called 'pear-shaped'; the fifth harmonic has five 'petals'; the seventh harmonic seven;

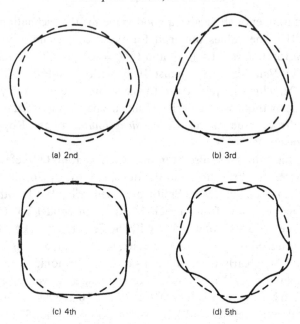

Fig. 3.8. Form of the second to fifth zonal harmonics. The sections shown are slices through the poles and the shapes give variations with latitude, averaged over all longitudes. The second harmonic corresponds to the Earth's flattening, the third expresses the 'pear-shape' effect, the fourth harmonic is square-shaped, the fifth has five 'petals', and so on.

and so on. The theoretical effects of odd zonal harmonics on orbits were calculated by Robin Merson and Rex Plimmer early in 1958; so, by the summer of 1958, we were well aware that odd harmonics would create an oscillation in the distance of the perigee from the Earth's centre. This cyclic variation in perigee distance would be in time with the movement of the perigee round the orbit. If the 'stem of the pear' was to the north, the perigee distance would usually be greatest when the perigee was furthest south ($\omega = 270°$) and least when the perigee was furthest north ($\omega = 90°$).

The perigee of Sputnik 2 moved at less than 0.5° per day, so the total movement during the lifetime was only 70° (from $\omega = 60°$ to $\omega = 350°$). Consequently the oscillation in perigee distance (now known to have had an amplitude of about 4 km) masqueraded as a nearly-linear change and was hidden in the much larger nearly-linear change (about 50 km) produced by air drag. We could not separate the two effects.

The US satellite Vanguard 1 suffered very little drag, and ω moved at 4° per day; so any oscillation in perigee distance dependent on ω should quickly show up. Sure enough, the NASA orbits of Vanguard 1 in the summer of 1958 revealed just such an oscillation, of amplitude about 4 km. But by now we had learnt too much about satellite orbits. In particular we knew that the NASA orbital elements might be either 'osculating' or 'mean': osculating elements are instantaneous; mean elements are averaged, usually over one revolution. If the elements were of the osculating type, there would be an oscillation of the kind observed, created merely by the definition of the orbital elements; if the elements were 'mean', the oscillation was revealing the effect of the odd harmonics. If we had been more ignorant, we should have leapt to this second conclusion. As it was, we cautiously avoided being the fools who rush in where angels fear to tread. 'Nothing venture, nothing win' would have been a better motto, however, because the oscillation *was* real, and was analysed by John O'Keefe and others to produce the first value for J_3, published in *Science* early in 1959, namely $10^6 J_3 = -2.4$. This value proved to be quite accurate; but the next chapter is the place for comparisons.

As Table 3.1 shows, the hopes of improving knowledge of the Earth's gravity field were being generously fulfilled. The value of $10^6 J_2$ previously accepted, 1091, had proved to be in error by 8, while the accuracies of the new values were beginning to approach 0.1 by the end of 1960. So, in three years of orbit analysis, the accuracy of J_2 had been improved by a factor of about 80, no mean feat. And several new coefficients, such as J_3 and J_4, which had never before been 'detected' observationally, were being evaluated with comparable accuracy.

The traditional ground-based geodesists had now ceased to oppose the new values: from 1960 onwards our work at the RAE was generously supported by the 'old-style' geodesists. They remained cautious about the possibility of further advances, but they warmly welcomed what had already been achieved. The 'new paradigm' had prevailed.

The dialogue between the traditional geodesists and the upstart space-people had started in June 1958 when the Royal Society's Geodesy Subcommittee set up a Study Group on Artificial Satellites, with Brigadier Guy Bomford as 'chairman and secretary' and Dr de Graaff-Hunter, Sir Harold Jeffreys and myself as the most active (or awkward?) members. Brigadier Bomford kept us going at a cracking pace by continually sending out letters with queries for individuals, and I have a file of hundreds of pages of documentation, much of it detailed mathematical analysis and queries. One of the most prescient ideas appears in an earlier letter dated 24 March 1958 from Sir Harold Jeffreys to Dr de Graaff-Hunter. The general gravity field, he says, needs to be expressed in terms of the associated Legendre functions $P_n^s(\sin\phi)$ of degree n and order s. Jeffreys then remarks that 'for n and s of order 15 there is a possibility of resonance, and the higher harmonics ... may have an appreciable effect'. As Chapter 5 will show, this insight was prophetic. But it was somewhat sabotaged by a further comment: 'on the whole I think that the only useful result likely to emerge is the check on the P_2 term [i.e. a value of J_2] and the possible correction for P_4 [i.e. a value for J_4].'

The Study Group sent in a long interim report on 5 December 1958, which included the numerical results up to then, the opinion that 'J_3 to J_6 or J_8 may ultimately be determinable from satellites with standard deviations of 1 in 10^6', and details of orbital theory developed by Merson and Plimmer, and by de Graaff-Hunter. After this, John O'Keefe became a corresponding member of the group, keeping us up to date with the US work. The final report of the Study Group, dated 1 January 1960, gives numerical results up to then, updates the orbital theory, and suggests a radar-altimetry satellite in a near-polar orbit to measure the geoid profile and a geosynchronous satellite to measure variations of the gravity field with longitude.

Another link with the traditional sciences was through the Royal Society's National Committee for the IGY (International Geophysical Year). The IGY of 1957–58 was an international enterprise covering all branches of Earth science and scheduled to last from July 1957 until December 1958. The satellites launched by the USSR and USA fitted into this research effort – a fortunate coincidence in timing that allowed the first satellites to fly as ambassadors of science rather than as military offshoots. The British National Committee for the IGY set up an Artificial Satellites Subcommittee in 1956, with Professor Massey as chairman, and I was appointed to this Subcommittee in April 1958. The Subcommittee had an 'Optical Methods Working Group', with Professor C. W. Allen as chairman and Mr P. Nuttall-Smith as the RAE representative. The Working Group was most useful in bringing out the scientific value of the kinetheodolite observations and in promoting volunteer visual observing. The Working Group produced a paper explaining the scientific value of visual observations, and in July 1958 this paper was sent out as part of an official letter from the Royal Society 'to persons interested' in visual observing. 'Your voluntary participation in the programme of observations outlined is cordially invited.' This invitation (IGY.CIRC/64(58)) began an enterprise that has lasted more than thirty years. The Royal Society still loans equipment to observers, advised by an 'Optical Tracking Working Group'. *Plus ça change* .... And some of the observers have had equipment on loan for more than twenty-five years.

When the IGY came to an end, the Royal Society set up instead a British National Committee for Space Research, with Professor Massey as chairman, to supervise British space research and act as the UK link with COSPAR. The National Committee gave birth to a subcommittee on Tracking and Data Recovery (TADREC), with J. A. Ratcliffe as chairman; this operated through four working groups, two of which were relevant to orbit analysis. One, with Professor Allen as chairman, was a continuation of the previous Optical Methods Working Group, with the name changed to 'Tracking Working Group'. This Group was at first much concerned with the kinetheodolites and radio tracking, and also in providing binoculars, stopwatches and star atlases for the volunteer visual observers who had responded to the Royal Society's cordial invitation. The second Working Group, with Dr Alan Cook as chairman, was on Orbit Analysis: it absorbed the activities of Brigadier Bomford's group, and several members of this group were 'carried over'. The Orbit Analysis Working Group first met on 4 June 1959: it included Dr de Graaff-Hunter, Sir Harold Jeffreys and myself, together with Robin Merson, Gerald Groves from University College, London, George Wilkins from the Royal Greenwich Observatory at Herstmonceux, and others.

The wider world

After swimming so long in the scientific fish-pond, I shall now try to struggle out. Although living mostly in the pond, I was also involved in 'public relations'. At first the RAE had been nervous about answering phone calls from journalists, but we soon had permission to do so. Consequently, our estimates of the likely lifetimes of decaying satellites, and even our impromptu opinions about new launches, often figured in the newspapers, and usually on the front page, because space was hot news in those days and not consigned to a small note on an inner page as it is today unless a disaster occurs.

Another time-consuming task was giving talks: the demand greatly exceeded the possible supply, and sometimes I wrote several letters a week politely declining invitations to speak. Because of the rapid pace of advance, some new slides usually had to be prepared for each new talk. The limit was set at one per month, with preference given to international conferences and Royal Society meetings rather than the Applestone Astronomical Society. In addition to the COSPAR meeting at Nice, I spoke at two other international conferences in 1960, the Aeronomy Symposium at Copenhagen already mentioned (July) and the Second

International Congress of the Aeronautical Sciences at Zurich in September. The flight to Copenhagen was via Amsterdam and Hamburg, so my acquaintance with continental European countries expanded from zero to five during 1960.

Editors of journals too were growling for articles. The most memorable growler was Dermot Morrah, who ran a highly-esteemed journal of current affairs called *The Round Table* and had the fruitiest voice I have ever heard. He was a friend of Sir Owen Wansbrough-Jones, who asked me to help Morrah in October 1957. Writing an article that might be intelligible to non-scientists proved a difficult task: it came out with the title 'A girdle round the Earth', and was mostly non-technical description of how satellites are launched and behave in orbit. By the summer of 1958, having written two further non-technical articles, I was not happy at the prospect of turning out more, for three reasons. The omnipotent editor chose the subject, whereas I would have preferred to make my own choice. The editor set a deadline, which might clash with 'real' emergencies, such as new launches. Worst of all, the editor would make alterations, so there was no incentive to try to write well.

My subconscious wish, I suppose, was to write a book, untrammelled by editors, about the researches made possible by satellites. With the good luck that seemed to cling like a burr in these years, that offer came in the autumn of 1958 from Norman Franklin, a director of the publishers Routledge and Kegan Paul, who knew of me because David Leslie was his brother-in-law. I accepted his offer and spent all my spare time in the next six months writing the book, which was entitled *Satellites and Scientific Research* and was published, after a delay due to a printing strike, in January 1960. By a curious coincidence, my book on Shelley was published by Macmillan almost simultaneously, under the title *Shelley: his Thought and Work*. My good-luck stories must be rather irritating by now, but it would falsify my perception of the events to pretend otherwise. Partly because of the simultaneity of publication, both books were favourably reviewed and sold well. Soon the *Shelley* was out in paperback, and *Satellites and Scientific Research* had to be revised for a second edition, published in 1962 in hardcover and paperback. The satellite book was also quickly translated into Russian on the initiative of Alla Massevitch, and proved popular in the USSR. Authors often suffer a nightmare feeling that no one actually reads their books right through: that feeling certainly afflicts me when I find the pages uncut in a Victorian book I have borrowed from the Cambridge University Library. Surely someone else must have read it in all those years? From this nightmare it is a short step to wondering whether all the work was worth while – for writing a book is always hard work. With *Satellites and Scientific Research* the answer was 'yes': the publishers had produced a good-looking book reasonably priced that sold out quickly and probably had quite a few readers. Later, Geoffrey Perry told me that *Satellites and Scientific Research* aroused his interest in space, which led to his many years of assiduous tracking and analysis of the USSR satellites from the Kettering school where he taught physics.

At the RAE in 1960 it was still 'full speed ahead'. The work on space research was well supported: a symptom of this support was my promotion in July to Senior Principal Scientific Officer in the 'individual merit' scheme, without any further administrative chores. My new salary was £3250 a year, a princely sum when you realize that my wife and I had recently bought a three-bedroom house for £2650 (and the mortgage interest rate was 4%). For me the 1960s were to be the decade of affluence, when a year's salary would buy a house. (After further promotion, followed by twenty years of annual erosion, my salary in 1987 was less than 20% of the current house price.)

4

Sailing through the sixties, 1961–1969

> Whither, O splendid ship, thy white sails crowding...
> That fearest nor sea rising, nor sky clouding,
> Whither away, fair rover, and what thy quest?
>   Robert Bridges, *A Passer By*

In 1961 a clear ocean of scientific research seemed to have opened up, ready to sail into and explore. The climate seemed set fair too. This optimism – fearing 'nor sea rising, nor sky clouding' – was justified by events: the 1960s proved to be a decade of fairly easy achievement, exploiting techniques already devised.

The RAE research on the upper atmosphere had so far been received in deafening silence by the Meteorological Office, which regarded anything at heights above about 20 km as rather 'way out' and of no interest to weather forecasters. This hardline attitude by meteorologists was slowly softening, and the Royal Meteorological Society invited me to give the Symons Memorial Lecture on 1 March 1961: the title was 'Satellites and the Earth's outer atmosphere', and I ranged more widely than in previous talks, discussing the history of ideas on the atmosphere and also venturing further outwards, above 1000 km height, into the exosphere and magnetosphere.

A month later came the most important scientific meeting I ever attended, the 1961 COSPAR Symposium at Florence. For this occasion we gathered all the data on air density for an updated picture of the variations with height, with solar activity and between day and night. Fig. 4.1 shows the graph of density versus height obtained from twenty-nine different satellites launched before 1961, as presented at Florence.[1] This graph was our first attempt to display both the decrease in density as solar activity declined between 1958 and 1961, and the changes between day and night.

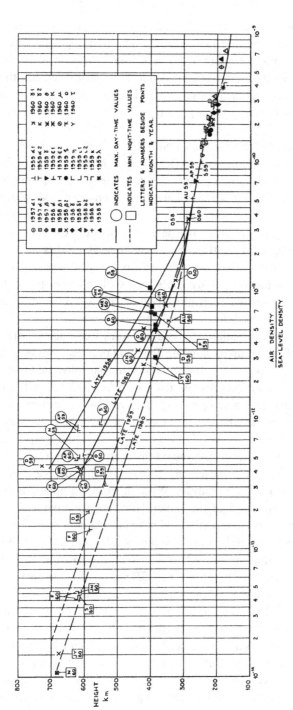

Fig. 4.1. Upper-atmosphere density obtained[1] from the orbits of twenty-nine satellites launched before January 1961.

At a height of 600 km the daytime density decreased by a factor of 3 between late 1958 and late 1960, and the minimum night-time density in late 1960 was about one eighth of the maximum daytime density. The day-to-night change was similar to that deduced a year earlier, but the great decrease in solar activity during 1960 had accentuated the solar-activity effect, which would presumably grow much greater as solar minimum approached.

Travelling to Florence for the COSPAR meeting, my first visit to Italy, was memorable in several ways. After flying to Milan, I went on by train to Florence: I was delighted by the Tuscan countryside, the antiquities at Florence and the Cascine woods where Shelley wrote his 'Ode to the West Wind'. Scientifically too there was much of interest at the meeting: Luigi Jacchia presented his analysis of the effects of the great solar storm in November 1960, showing that the air density at heights near 600 km jumped up for a few hours by as much as 8 times in response to the atmospheric heating caused by the streams of particles from the Sun impinging on the Earth's outer atmosphere. This was a beautiful piece of work, recognized as such at the time, and ever since.

Most fruitful scientifically for my future work were the meetings of the COSPAR working group on Satellite Tracking presided over by Professor Fred Whipple, the Director of the Smithsonian Astrophysical Observatory (SAO) at Cambridge, Massachusetts. The SAO had taken the lead in satellite tracking and was operating the world-wide system of twelve Baker–Nunn cameras that were capable of recording accurately the positions in the sky of very faint satellites, even the grapefruit-sized Vanguard 1. The SAO also determined the orbits of the satellites observed by the cameras, and was now leading the world in this role – and providing rich material for Jacchia's analyses. Without Fred Whipple, it is doubtful whether the construction of the Baker–Nunn cameras would have been pushed forward: certainly, they would have been later into service, and the progress of world research in orbit studies would have been much slower.

Equally important in the COSPAR Working Group was Dr Alla Massevitch, who was in charge of the USSR satellite tracking network, and was soon to chair the Working Group in succession to Fred Whipple. These two were both determined that the satellite observations made throughout the world should be coordinated, and available to scientists of all countries. This was no easy task at a time when Cold-War phraseology was fashionable and the Cuban missile crisis was yet to come. But the generous idea of Whipple and Massevitch was unanimously supported, and the international arrangements agreed at Florence, for coding observations and revealing the positions of observing sites, stood the test of time. Without this international goodwill, our future work on satellite orbit determination would have been strangled at birth, because the calculation of unbiased and accurate orbits depends on a wide geographical spread of observations.

Half-way through the COSPAR Symposium, on the morning of 12 April 1961, the first man to fly in orbit round the Earth, Yuri Gagarin, made his historic journey. Alla Massevitch invited me to a celebratory lunch in Florence that day with two other scientists from the conference. Silly as it now seems, a general euphoria prevailed in 1961: as the scientists met in the splendid Palazzo Pitti, they half-felt they might somehow move heaven and earth by sending robot and human explorers into space. It is difficult now even to remember that feeling.

Returning from the Palazzo Pitti to the non-palazzos of Farnborough was down to earth indeed, and all the more so because Doreen Walker left in April and was to be away for seven years bringing up her family. Her successor at the RAE, Janice Rees, had to 'start from scratch', but was soon hard at work on calculating air densities.

Larger-scale changes at the RAE were also in the mill. In January 1962 James Lighthill created a Space Department, which was allowed to steal more than half the staff from the existing (but now misnamed) Guided Weapons Department. The first Head of the new Department was Dr A. W. Lines, and I was in the Dynamics Division, then headed by George Burt. These changes were most helpful: we were now officially working on space research and not renegades slinking away from guided weapons. It is some measure of the optimism of these days that I suggested to James Lighthill a change in the name of the RAE to 'Royal Aerospace Establishment'. That idea was vetoed by the strong 'aircraft' side of the Establishment. The name was eventually adopted twenty-five years later, on 3 May 1988, which was, ironically, the day I retired. But that was far ahead.

The COSPAR meeting at Florence was quickly followed by another space conference, at Paris in July. This was, I believe, the first scientific meeting to be organized by the newly-formed International Academy of Astronautics. The guiding spirit behind the Academy (and the conference) was the eminent aerodynamicist and rocket pioneer, Theodore von Kármán, who had worked in California since emigrating from Hungary in the 1930s. The subject of the conference was 'the trajectories of space vehicles', and I spoke of our work on the effect of atmospheric oblateness on orbits.

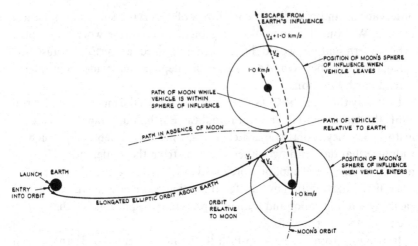

Fig. 4.2. Diagram showing how the Moon's attraction can be exploited to enable a spacecraft to escape from the Earth's influence. The spacecraft needs to pass just behind the Moon to achieve the maximum effect – an additional velocity of 1.0 km/s.

But a more important RAE contribution to the conference came from Harry Hiller, whom I have not yet mentioned. He began working with me on the missile design studies in 1953, but was not immediately brought into the satellite orbital studies in 1957. However, he was allowed to 'go into space' in 1958, going beyond satellites and concentrating on space vehicles that escape from the Earth's influence altogether: they usually have 'open' hyperbolic orbits rather than the closed elliptic orbits of satellites. Harry Hiller began his deep-space work with a lengthy analytical study of the trajectories of spacecraft travelling from the Earth into the sphere of influence of the Moon. At the Paris conference he presented a summary of this work, which von Kármán commended. Hiller's paper[2] had many interesting facets, but in retrospect the most significant finding was that, 'for vehicles passing behind the Moon, there is always a gain in total energy of the vehicle at the expense of the Moon' as the vehicle crosses the Moon's sphere of influence (taken to be of radius 41 000 miles). As seen from the Earth, the spacecraft speeds up, and the Moon slows down infinitesimally. The spacecraft emerges from the Moon's sphere of influence with a greater velocity (relative to the Earth) than when it entered, and the gain is greatest if the spacecraft skims the right-hand rim of the Moon. This was probably the earliest full mathematical study of the use of the Moon (or planets) to accelerate passing spacecraft, a technique that has been widely used, for example with the Voyager and Galileo spacecraft. The idea of these 'gravity-assist' trajectories is usually credited to later US scientists, but Hiller's work deserves to be recognized too, and so does an earlier analysis by V. Yegorov.[3] Fig. 4.2 shows that the increase in velocity produced by skimming the Moon can be up to 1 km/s.

Later in 1961 I was elected a Member of the fledgling International Academy of Astronautics, which has now grown into quite a large organization devoted to keeping up standards in astronautics.

After lingering so long in 1961, I shall now pursue the various strands of research right through the 1960s, rather than taking events year by year, so as to preserve some semblance of continuity in discussing each research topic.

Determining satellite orbits from observations

The special campaign of kinetheodolite observations of Sputnik 2 had enabled Merson to derive better orbits than anyone else in the world. This was the key to the research successes in 1958–59, but the observations would not continue through the 1960s.

Such accurate orbits were not needed, however, to calculate air density from orbital changes. The orbital period must be accurately known, but an error of, say, 2 km in height would not be important. In 1960, we and most other upper-atmosphere researchers relied on the orbits kindly supplied by the 'Space Track Control Center' of the US Air Force, which had the task of keeping track of all objects in space and maintained orbits of all the satellites for prediction purposes. These orbits, which were made available to researchers, were not of high accuracy, and errors of up to 5 km in perigee height could be expected. But the orbital period T was usually very accurate, and could be checked by comparing the changes in the predicted times between the end of one bulletin and the beginning of a new one. So we were content to calculate orbital decay rates from these 'Spacetrack orbits', usually by taking two values of T a few days apart and dividing the difference by the time interval.

This source of orbital information continued throughout the 1960s: indeed it still continues today. In 1962 the North American Air Defense Command (NORAD) was formed, and for the rest of the decade NORAD was the source of the data used in many determinations of air density.

More accurate orbital information on a number of satellites was supplied by the Smithsonian Astrophysical Observatory. The SAO determined orbits, and issued predictions based on these orbits, for the twenty or so satellites being observed by their Baker–Nunn cameras. This service was maintained throughout the 1960s and was valuable as a source of higher-quality data for these satellites. Of even higher quality were the orbits determined by the SAO in retrospect: these orbits were published in the series of SAO Special Reports, of which 200 were issued by 1966. More than half of these Reports related to satellite observations, orbits or analysis, and were a real treasure-house of information. The data covered about fifty satellites, sometimes over the whole lifetime, sometimes only for a few months.

A third source of orbital data was the US National Aeronautics and Space Administration (NASA). Nearly all satellites launched by NASA carried radio transmitters operating at a frequency near 136 megahertz. These radio signals were received by a network of twelve 'Minitrack' radio interferometers round the world, including one in England at Winkfield near Ascot, set up in 1960 and operated by the Radio Research Station at Datchet (which was of course also the UK prediction centre). With the aid of the Minitrack observations, which were accurate to about 0.02° in direction, NASA determined the orbits of their satellites and distributed the information to interested scientific organizations. The observations (in the form of direction cosines) were also made available – one example of the generous distribution of data by NASA, so helpful for all types of space research.

With these three sources of orbital data, you might think we would have been able to analyse orbits very well without any 'home-grown' orbit determination. But that was not so. The Spacetrack (later NORAD) orbits were not usually accurate enough for any purpose other than evaluating air density. The SAO determined only the orbits of satellites observed by the Baker–Nunn cameras: these were nearly all American and nearly all were in orbits little affected by the atmosphere (though there were exceptions, such as the two Echo balloons). The NASA satellites were also usually launched into high orbits uninteresting for atmospheric research. The most useful orbits for analysis were those of the Russian Cosmos satellites and final-stage rockets in low orbits with lifetimes of between 6 months and 3 years; and for these it was obvious that we should have to determine our own orbits if we wanted to extract the information lying latent and waiting to be teased out.

Robin Merson was already skilled in determining orbits from observations: he now developed for use on the Pegasus computer a new program that would accept observations of several types. The previous program had determined the orbit from many observations on one transit, as from a kinetheodolite. The future trend would be for few or single observations from geographically separated areas.

A vital preliminary for the program was the proper definition of the orbital elements (*a*, *e*, *i*, etc.) for an orbit in the gravity field defined by the series of harmonics J_2, J_3, J_4, ... mentioned on p. 40. The orbit in this real gravity field is no longer an exact ellipse, and the instantaneous ('osculating') values of the elements vary slightly as the satellite proceeds round the orbit. 'Mean' elements, averaged round the orbit, were therefore needed. But there were various possible ways of averaging out the variations (indeed we did not know until the late 1960s the exact definitions of the elements produced by NORAD and NASA). So Merson began by calculating the effects of the individual J_n on each of the orbital elements: the work was published[4] in 1961, and later in revised form as RAE Technical Note Space 26 (1963).

Merson's results helped to define mean elements for the new computer program, which was ready in time to determine the orbit of Ariel 1, the first Anglo-US satellite, launched in April 1962. The Minitrack observations of Ariel 1 were used to calculate orbits at intervals of twenty-five revolutions (approximately 2 days) from April to August 1962. The orbital elements proved to be accurate to about 100 m cross-track – better than had been expected, and a portent of future possibilities.

The Pegasus computer was soon to be superseded by much more powerful machines, which offered the chance of developing a new and more flexible program for the mid 1960s. This was to be called PROP, standing for 'Program for the Refinement of Orbital Parameters', the 'refinement' being produced by successive iterations and the selective rejection of ill-fitting observations. PROP was to serve as the work-horse of orbit determination in the UK for more than twenty years – indeed it is still in use. The orbital model for PROP was defined by Merson in 1966. But by this time he was handing over the work on determining the orbits of low-altitude satellites to Bob Gooding, who had come to Farnborough in 1958 after working with the kinetheodolite at Orfordness, and soon became interested in orbit theory. His doctoral thesis on 'satellite motion in an axisymmetric field' was issued in 1966 as RAE Technical Report 66018, and from then on he continued to work on orbit theory and determination, creating a sound basis for all our orbit analyses.

The first orbit determination with PROP was done by Bob Gooding in 1967 (with London University's Atlas computer), using NASA Minitrack observations of the satellite Secor 6. The program became generally available for use with the RAE's ICL computers in 1968 with the issue of the Users' Handbook (RAE Technical Report 68299). From then onwards, those of us who analysed orbits became PROP addicts, or so it seemed; it was our daily companion, and the work would have lapsed into mediocrity without it. If we had problems, Bob was always ready to explain and help us, and much of our scientific success in the 1970s and 1980s depended on his far-sighted programming.

The question of how to derive accurate orbits was thus in process of being answered during the 1960s: however, there was still the separate question of how to decide in advance which orbits are best for observing and analysing. It was to help in answering this question that the RAE Table of satellites was kept going during the 1960s, and beyond. In the mid 1960s it used to be issued once a month to about 200 people, including about 60 visual observers, and this monthly production and distribution (on a smaller scale) has continued into the 1990s. By the end of 1969 the Table extended to 215 pages and was growing by about 30 pages per year. We were greatly helped in producing the Table in the 1960s by Alan Pilkington, who had worked in the prediction service at the Radio Research Station in the early 1960s and later ran his own Planetarium at Scarborough.

From the orbits and lifetimes in the Table, likely satellites for observation and analysis could be selected. The lifetimes needed to be reasonably long to obtain enough orbits for analysis, but not too long if a whole-life analysis was in prospect. The satellites also needed to be bright enough to be seen by optical observers: the size and shape recorded in the Table gave an idea of the brightness. The correct choice of the 'seed-corn', for harvest in our analyses, was an important matter: if we got it wrong, much effort would be wasted. Thanks to the Table, this rarely happened.

Observations

Between 1961 and 1967, while the programs for determining orbits from observations were maturing, the observations available were also changing, sometimes for the worse but mostly for the better.

The observation of Sputniks 2 and 3 by RAE kinetheodolites was a special project of limited life. In 1960 the RAE ended its operational involvement, but agreed to make available three kinetheodolites on extended loan for satellite observing by other agencies. The first of these was installed at the Meteorological Office station at Qrendi, Malta, and thousands of observations were made there between 1960 and 1968. The second went to the Royal Observatory Edinburgh and was used extensively between 1961 and 1969, but little thereafter. The third kinetheodolite was set up, after some delay, at the Royal Greenwich Observatory,

Herstmonceux, and was productive between 1964 and 1969. So, by the time the PROP orbit determination program got going in 1968, plenty of kinetheodolite observations were waiting to be used.

Many other observations were available from theodolites, at Jokioinen in Finland, Meudon in France, and elsewhere; from large cameras such as the twelve Baker–Nunns; from smaller cameras, such as the Russian NAFA 3c/25, the US Minitrack Optical Tracking System (MOTS), and others. There were also the Minitrack radio observations already mentioned. Unfortunately, most of these instruments were tracking satellites of interest for particular projects – usually because their orbits needed to be known so as to interpret the results of on-board experiments. The orbits were rarely of much interest for orbit analysis. The exception to this rule was the Jokioinen theodolite,[5] which kept up observations on satellites useful for orbit analysis for twenty years: these observations were very valuable in orbit determination because of the high latitude of the station. The angular accuracy of the kinetheodolite observations was 0.01° at best; with the theodolites, the figure was usually between 0.03° and 0.05°.

The splendid prospect of much better accuracies was emerging in 1961 with the birth of Hewitt cameras. These were two large cameras designed in the late 1950s by Joseph Hewitt (1912–75) of the Royal Radar Establishment (RRE), Malvern, with the aim of recording photographically the luminous atmospheric entry of the Blue Streak ballistic missile. By the time Blue Streak was cancelled, two magnificent cameras had been manufactured by Grubb Parsons. Now they were looking for a role. Joe Hewitt was allowed to adapt them for satellite tracking, and in 1961 one camera was set up at an excellent site at Sheriff's Lench, near Evesham. (The other was sited at Lye Ballets in Herefordshire, but was never used there.) The camera at Evesham, often called the Malvern camera, was operated on an experimental basis by Joe Hewitt and his colleagues between 1962 and 1967; there was no regular programme for satellite observing at that time, but the camera contributed to some of the SAO observing campaigns, to supplement the Baker–Nunn cameras and to assess its accuracy.

The Hewitt camera soon showed itself to be quite superb, both optically and mechanically: it became, and remained, the most accurate satellite camera in the world, capable of accuracies of about 1 second of arc (0.0003°), though often used in a less accurate mode (2–3 seconds of arc) to save time in reading the photographic plates.

The Hewitt camera, Fig. 4.3, is of field-flattened super-Schmidt type, with an aperture of 61 cm (24 inches). Its field of view is 10° in diameter.

Fig. 4.3. The Hewitt camera, designed and developed at the Royal Radar Establishment, Malvern, by Joseph Hewitt. The camera has an aperture of 61 cm and can photograph satellites with a directional accuracy of 1 second of arc. This is the camera which operated at Siding Spring Observatory in Australia from 1982 until 1990.

The camera remains stationary and records the track of the satellite as it crosses the field of view, the track being broken at known times by a shutter. The faintest satellite that can be recorded depends on the angular rate of travel across the sky. A fast low satellite, moving at 1° per second, needs to be of stellar magnitude 7 or brighter. A satellite at a distance of 4000 km, moving at 0.1° per second, can be recorded if it is brighter than magnitude 10, while geostationary satellites can be recorded down to magnitude 15.

The layout of the Hewitt camera is shown in Fig. 4.4. The light enters from the left through the Schmidt corrector plate and is focused by the 86 cm diameter mirror on to the photographic plate. A lens in front of this plate ensures that the rays converge on a plane rather than a curved surface. The camera can be operated in many different modes, depending on the likely behaviour of the satellite. The accuracy of about 0.0003° was about 3 times better than the Baker–Nunn camera, which relied on film stretched over a curved surface and therefore liable to distortion.

In 1967 the two Hewitt cameras were purchased from the RRE by the Ordnance Survey, on the initiative of the Director-General, Major-General

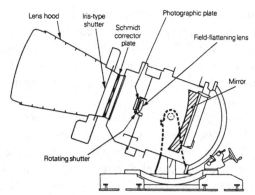

Fig. 4.4. Cross-section of the Hewitt camera. When a satellite is being photo-graphed, the iris shutter is open and light passes through the corrector plate to the mirror, where it is reflected and passes through the field-flattening lens on to the photographic plate (or film fixed to a glass plate).

R. C. A. Edge. The first camera remained at Sheriff's Lench, Evesham; the second went to the Royal Observatory Edinburgh. The observations made by the cameras greatly improved the accuracy of our orbits in the 1970s, as Chapter 5 will show.

At the opposite extreme to the Hewitt camera were the homespun observations made by the visual observers with their binoculars and stopwatches, often observing from a deck-chair in the garden. These (myself included) were the volunteers 'cordially invited' to offer their services by the Royal Society in 1958. The response was more than cordial: indeed, to adapt an old saw, satellite observing sometimes seemed to be 'the cordial drop in the otherwise vapid cup of life', and by the mid 1960s there were about 50 skilled observers making observations of some 20–40 satellites being predicted by the Radio and Space Research Station (RSRS), as the RRS was by then known. Most of the observers were in Britain, but others came from far afield, from Fiji and South Africa, from New Zealand and the USA, as well as from Europe. The observers were supplied with the equipment they needed – chiefly binoculars, stopwatches and star atlases – on the recommendation of Professor Allen's Tracking Working Group, which in 1966 was transformed into the Optical Tracking Subcommittee of the British National Committee for Space Research. Some new members were added in 1966, including Russell Eberst, by then the world's leading visual observer. The Royal Society, which organized the meetings of the Optical Tracking Subcommittee, stored and loaned the equipment (and still does so). The prediction service at RSRS was headed by David Smith during much of the 1960s. Before he left for the USA in

1968, he recruited a keen French observer, Pierre Neirinck (who gradually took over more responsibility for the predictions in the 1970s). The satellites predicted by RSRS were those selected by the Subcommittee as most likely to be useful for orbit analysis. Most were bright satellites strongly affected by drag, the rockets of Russian Cosmos satellites being particularly favoured.

Professor Allen's Subcommittee also arranged occasional meetings specially for observers, where the value of the observations could be explained and observing techniques could be discussed. The first two such meetings were held at the Royal Society in 1965 and 1968. The 1968 meeting included some interesting experiments, devised by Gordon Taylor, to assess the positional and timing accuracies achieved by observers. The average positional error was found to be about 1/20 of the distance between the reference stars; the average timing error was 0.08 s, appreciably less than had been expected.

What do you actually have to do, to achieve such accuracy in visual observation? The predictions tell you approximately the track of the satellite across the sky, and the time, accurate typically to about one minute or better for satellites of fairly high drag. About two minutes before the predicted time (or more if you suspect the predictions may be in error), you go to a suitable place for observing, perhaps (Fig. 4.5) setting down a deck-chair at some place in the garden untroubled by moonlight, street lights, or house lights, which have to be turned off or curtained. (Observing from indoors through a window is possible but not recommended.) Then you look through the binoculars towards a convenient point on the selected track, see the moving satellite within a minute or so, if fortune is smiling, and follow it until it passes two convenient stars, preferably stars that are close together and perpendicular to the track of the satellite. As the satellite crosses the (imaginary) line between the two stars, you estimate the crossing-point – say, 7/10 up from star A to star B – and at the same time start the stopwatch, which is later stopped against a time signal. With practice, two more such observations further along the track can be made.

Observing is a skilful art: to begin with, the prediction must be interpreted correctly, with judicious allowance for unknown errors in the timing; the 'pick-up point' on the track needs to be well chosen and quickly identified in the sky; the reference stars, often picked 'on the wing' rather than beforehand, need to be close together and correctly identified; the timing of the observation must be exact to the limits of human capability; and so on. By the mid 1960s, skilled observers were achieving accuracies of 0.03° in direction by using reference stars less than 0.5° apart,

Fig. 4.5. Visual observer restfully at work. Note the mobile observatory (deck-chair), the hood and the screen of evergreen trees (both to avoid lights and cold winds). A third hand for holding the stopwatch would be useful (and a fourth for a second stopwatch).

and 0.1 s in time (as the 1968 experiment showed): these observations were quite accurate enough to be of value in determining orbits, especially for satellites of fairly high drag.

In the 1960s many British visual observers also sent their observations to the Moonwatch organization at the Smithsonian Astrophysical Observ-atory, which maintained close links with the UK prediction service. Moonwatch was set up in 1957 to encourage world-wide visual observa-tions of the satellites of interest to the SAO. The observations were intended primarily for updating and improving the predictions, and thus raising the success rate of the Baker–Nunn cameras. When the SAO determined the orbits of these satellites, the visual observations were mixed in with the Baker–Nunn observations. As the orbit was accurately known, the errors in the visual observations could be assessed. Moonwatch sent visual observers the 'residuals' of their observations (in effect their errors), and this service was much appreciated by the observers, as nothing similar was available in the UK until after 1968. When they knew of the errors, many observers thought they could do better; they tried to do so, and usually succeeded. The heyday of Moonwatch was in the late 1960s, when the Head of Moonwatch, Bill Hirst, had funds for travelling widely to visit and encourage the observers, a psychological boost to keep up the enthusiasm of the volunteers in their lonely and exacting measurements. There were also numerous visual observers in the USSR, separately organized by the Astronomical Council of the Soviet Academy of Sciences: some of these observers were just as keen as those in Britain, and their observations were also used in determining orbits.

Satellite observing was a new subject in the early 1960s, and there were no books about it. Macmillan published a book of mine on Erasmus Darwin in 1963, as well as the *Shelley* earlier, and welcomed the idea of a book on observing. I was not unqualified for the task, because I had made more than 3000 visual observations by 1966 and knew about the scientific uses of the observations. I was rather ignorant of other aspects, but learnt a lot more while writing the book, which was entitled *Observing Earth Satellites* and published in 1966. The sales were good, but not many new satellite observers appeared: the deterrent seems to be the considerable skill required of a good observer. One new observer who was drawn in through reading the book was David Hopkins, whose observing feats subsequently grew to rival those of Russell Eberst.

In satellite observing, as in many other ways, the 1960s now shine as a golden era. People were interested in seeing satellites hurrying across the sky, and this interest was heightened by the launch of the bright Echo 1 balloon in 1960, and then by its even brighter successor Echo 2, 41 m in diameter, in 1964. These balloons were at greater heights than Sputnik 1 and took longer to go from horizon to horizon: they appeared as first-magnitude stars trundling slowly across the sky for about five minutes. Echo 2 was probably seen by more people than any other man-made object in the history of the world and, even though most of those seeing it would not have known what it was, this 'companion in the sky', with its two acres of reflecting aluminized plastic seen night after night, was beginning to create a new astral culture – until it decayed in 1969. Another large balloon called Pageos (*pa*ssive *geo*detic *s*atellite) was launched in 1966 into a higher orbit at 4000 km height. This was fainter than the Echos, but good for simultaneous observation by cameras in different countries (including the Hewitt cameras).

Large balloon-satellites, of diameter 40 m or more, are unrivalled for boosting public interest in space: the cost of launch is trivial by comparison with a Shuttle launch – which often leaves nothing in orbit. Yet none has been launched for twenty-five years: I tried to have Black Arrow used for this purpose, but in vain. Why the reluctance? My answer to this conundrum is that too cheap a project, offering no substantial profits for industrial companies, does not generate the lobbying that impresses administrators. They prefer an expensive project strongly supported by industrial lobbyists and also having the merit (for them) of needing a substantial administrative framework. The same problem has bedevilled orbit analysis: the body of scientific knowledge is enriched but nobody else is, so no one of consequence lobbies for it – only ivory-tower scientists.

The main advances in observing in the 1960s were optical and visual, but radio tracking still continued, with the NASA Minitrack system in full operation. Various military radars were also being devoted to satellite tracking, and the most important of these in Britain was the 45 ft dish at the Royal Radar Establishment, Malvern. With a staff of about ten this excellent instrument was very productive throughout the 1960s, and made thousands of observations of satellites suitable for orbit analysis, in addition to its military commitments in support of NORAD.

In the early 1960s NORAD was setting up a world-wide system of radars for early warning of ballistic missile attack. One of these stations, built in 1962, was in England, on Fylingdales Moor in Yorkshire. The primary aim was to detect ballistic missiles, but the radars also detected any satellites that were large enough and close enough. In the early years these observations were not generally available, but that changed in the late 1960s, and the observations from Fylingdales were extremely useful from then onwards.

Before this lightning survey of satellite tracking in the 1960s comes to an end, there is another source of data that must be mentioned, the observations made by the US Navy's Navspasur system. These, and the Hewitt camera observations, were to be the key to most of our successes in the 1970s. However, we did not use either type of observation until 1969; so I shall shunt them into the next chapter, and now turn to the research of the 1960s.

Air density and its variation with height

The results presented at Florence in 1961 have been shown in Fig. 4.1. After that, there was quite a long interval before we drew up the next diagram of this type, which appeared[6] in September 1963 and covered the changes in air density over the years 1957–63.

Earlier in 1963 I had been asked to speak at a conference to be held in August at the University of West Virginia, Blacksburg. As must be obvious by now, I am not an avid traveller, but this invitation was tempting because it included first-class air travel and a fee almost equal to a month's salary. The transatlantic flight by Boeing 707 was so luxurious that it seems almost phantasmagoric by comparison with the sardine-tin airliners of today. Back down on American earth, on the night drive to Blacksburg from Roanoke, I sampled skunks (olfactorially); and the University campus was alive with cicadas compulsively chirping out one of their seven-year plagues. At the meeting I presented the new picture of air density.[6] Tropical thunderstorms enlivened the return flight from Roanoke to Washington, where the temperature and humidity were both in the high nineties. I spent the rest of the day in the air-conditioned Library of Congress, admiring the speed with which the librarians produced eighteenth-century books for their unannounced visitor.

Our next (and last) general picture of air density[7] was published in 1965 and covered the half-cycle of solar activity, from sunspot maximum in 1958 to the minimum in 1964. This picture, reproduced here as Fig. 4.6, relies on values from 46 satellites; the variations of density with solar activity and between day and night are shown by the curves drawn through the many points. At a height of 500 km the daytime density was 10 times greater in 1959 than in 1964. As the density in 1964 was 4 times greater by day than at night at 500 km height, the daytime-1959 density was 40 times greater than the night-time density in 1964. At 600 km height this factor had grown from 40 to about 100. The upper atmosphere was proving a very volatile creature indeed, ever ready to delude a superficial enquirer who failed to specify the year and the time of day.

Janice Rees was co-author of the 1963 paper, but she left in that year, and was succeeded by Eileen Quinn, who was co-author of the 1965 paper. Another valued new member of the group, who joined in 1964, was Diana Scott, a skilled computer programmer. She worked mainly with Graham Cook on studies of the upper atmosphere and gravity field.

There were two further papers in 1965–66 on the variation of density with height at the lower end of the height range, 150–300 km. But, because of the clutter of points, there were no further comprehensive diagrams like Fig. 4.6: they had served their purpose by clearly revealing the huge variations in density and alerting scientists in related areas to the complexity of the upper atmosphere. From now on, the way ahead in upper-atmosphere research was to fit 'model atmospheres' to the observational data. This was the aim of the COSPAR Working Group on the atmosphere, of which I was a member during the 1960s. The idea was to define a 'reference atmosphere', so that new results could be compared

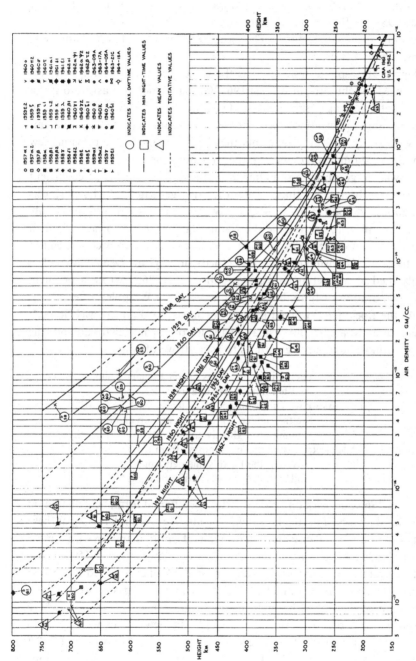

Fig. 4.6. Upper-atmosphere density during 1958–64, obtained from the orbits of 46 satellites.[7] Letters and numbers beside points indicate month and year.

with this standard: its defects could then be exposed and corrected. The first COSPAR International Reference Atmosphere came out in 1961, but was not very satisfactory. The second appeared in 1965 and was considerably better.

Upper-atmosphere research was the subject of a symposium at Cambridge, Massachusetts, in August 1965. Graham Cook was asked to speak about satellite drag coefficients, and I was invited to review our methods of determining upper-atmosphere density. Graham reported on the impressive study of drag coefficients that he had made in the previous few months. Although some uncertainties remained, he was able to show that there was no good reason to alter the value of 2.2 for the drag coefficient that we had been using since 1959 for low-perigee satellites. His work was not improved upon in the subsequent twenty years, as it turned out. I duly gave my talk too, but for me the Cambridge meeting also had a hidden agenda, because Fred Whipple had recently invited me to become head of the theoretical division of the SAO in succession to Imre Izsak, who had so sadly died at a meeting in Paris in April, aged 36. I wrote a poem in commemoration of Imre, but did not accept Fred Whipple's offer.

As the Boeing 707 winged its way north-eastward on the way home, I was treated to a splendid display of the aurora near Newfoundland, the upper atmosphere playing with its treasure-trove of colours:

> Each minute ten miles nearer England's shore
> And ten times wiser in auroral lore,
> As we watch the airy curtains flicker back and forth,
> See the sudden searchlights stab up and die,
> Column after ghostly column balanced in the sky.
> Pale electric atom-streams shooting from the Sun
> Have felt the Earth's magnetic might
> And spiralled in to beautify the night.[8]

After 1965 our work on air density took a different form, being based on detailed analyses of particular satellite orbits. The most interesting of these was Cannonball, a dense spherical satellite 0.6 m in diameter with a mass of 272 kg, launched in July 1968. Despite a very low perigee of 150 km, the satellite remained in orbit for 38 days. Doreen Walker returned to work in April 1968, and this was an ideal satellite to analyse: it gave a well-defined variation of density with height in the height-range 125–180 km, where information had previously been scarce.[9] The new values were close to those in the 1965 COSPAR International Reference Atmosphere (*CIRA 1965*) and thus validated *CIRA* at heights below the lower limit of previous results from satellites.

We also analysed the orbit of a satellite launched with Cannonball, called 'Spades', which added to these low-altitude results. Although we did not realize it at the time, our results appeared before those of the US scientists for whom the satellites were launched. This caused some ill-feeling towards us. But it is difficult to define the offence we had unknowingly committed, as our data were independent of theirs and the idea that anyone is allowed to observe and analyse a satellite in orbit had become well-established *de facto* after the precedent of Sputnik 1.

The years 1961–69 had seen a rich harvest of results on air density. In 1960 the only values available were for high solar activity. By 1969 a much better picture had emerged, covering a full eleven-year cycle of solar activity, and with the height limits greatly extended.

Variations of air density with time

Fig. 3.4 has shown how the air density increases and decreases in unison with the solar 10.7 cm radiation (which is a measure of the extreme ultra-violet radiation from the Sun) and with geomagnetic storms (caused by streams of particles from the Sun). This subject was being studied in detail at SAO by Luigi Jacchia, who had the best facilities, being able to use observations from the Baker–Nunn cameras. So the systematic work was ceded to him, although of course we often came across examples of these effects in analyses of particular satellites.

Instead we began looking at another curious variation of air density with time, which took on greater importance in the years of low solar activity, 1963–65. This is the variation loosely known as 'semi-annual', though it really has both semi-annual (6-month) and annual (12-month) components: it is also extremely irregular in form and strength, just to add to the confusion. Discovered in 1960 by H. K. Paetzold, the semi-annual variation was at first not easily separated from the dominant effects of the variable solar activity. By 1963, however, the accepted view was that 'upper-atmosphere density undergoes a semi-annual variation with minima in January and July, and maxima in April and October', to quote from a paper[10] presented in September 1963. It was already apparent by then that the July minimum was deeper than that in January, and that the October maximum was usually somewhat higher than that in April. As solar activity declined between 1962 and 1965, the semi-annual variation stood out more clearly.

Our first study of the semi-annual variation, in 1966, was provoked by a claim that the effect was really a variation of density with latitude. This idea was refuted by analysing a polar satellite in a nearly circular orbit which sampled all latitudes almost equally. It still showed a semi-annual variation.

At heights near 200 km the semi-annual variation in 1966–67 proved to be more important than the variations due to solar activity. This conclusion was reached[11] after deriving 154 values of density between June 1966 and July 1967 from the orbit of Secor 6: the density was about 40% greater in October 1966 and April 1967 than in July 1966 and January 1967.

Monitoring of the semi-annual variation continued with the aid of the satellite 1966–118A in a circular orbit at a height near 500 km. This work led to the suggestion of a 33-month periodicity in the amplitude of the effect between 1958 and 1968, possibly linked with the 'quasi-biennial' oscillation in stratospheric winds. 'If this correlation proves to be valid', we remarked, 'it may provide an interesting link between the upper and lower atmosphere'.[12] In fact, signs of the correlation have also been visible in later years, but not consistently enough to provide a useful method of prediction. It remains a tantalizing possibility.

Graham Cook and Diana Scott were studying the effects of the semi-annual variation at much greater heights, near 1100 km, chiefly by analysis[13] of the balloon-satellite Echo 2, and also at 500 km, by analysis of several satellites. In 1969 Graham Cook wrote a definitive review-paper on the semi-annual variation,[14] showing how the effect varied from year to year during the 1960s, and also how it depended on height for various levels of solar activity. For example, in 1967 the October maximum density exceeded the July minimum density by a factor that increased from 1.5 at 250 km to 3.0 at 500 km (and possibly 4.0 at 700 km) and then decreased to 2.5 at 1100 km. The April density was usually a little lower than that in October, and the July minimum was deeper than that in January. These were huge calendar-dependent variations that needed to be calculated and set in a rational framework in order to make sense of other atmospheric variations. The multiplicity of the variations was daunting, but their correlation with separate external parameters – such as solar activity, darkness and light, and month of the year – was a welcome step forward in understanding.

Many of these results were presented at international conferences, particularly the annual COSPAR meetings. I went to the 1966 COSPAR meeting, held in the imperial splendour of the Royal Palace in Vienna, and to the 1967 meeting in the unlovely lecture-rooms of Imperial College, London. In August 1968 I found myself in Vienna again, for a United

Nations conference on the peaceful uses of space, and as I flew back over Czechoslovakia the Soviet troops were entering. The 1969 COSPAR meeting was held at the International Hotel in Prague, a year after that 'chill wind from the east', in the phrase of a Czech colleague, Milan Bursa, who met Graham Cook and me at the airport. The hospitality was overwhelming: we were enjoying a tour of the dungeons in Prague Castle little more than an hour before the plane was due to leave, but were whisked to the airport in time.

Determining the density scale height

As explained in Chapter 3, the decrease in the perigee distance of a satellite as time goes on depends directly on the value of the density scale height H, being given approximately by $\frac{1}{2}H\ln(e_0/e)$. In an ideal spherical world, values of H could be found from the measured decrease in perigee distance. For orbits round the real Earth, however, there are quite large changes in perigee distance (of up to about 10 km) due to the effects of odd zonal harmonics, and smaller changes (not usually more than 1 km for orbits close to the Earth) due to the gravitational attractions of the Sun and Moon. Both these effects must be carefully removed, and the orbits must be accurate and reliable, if the rather small decreases in perigee distance due to drag, typically about 5 km, are to be used to evaluate H. Also the effect of atmospheric oblateness needs to be removed, and the theory must allow for the increase of H with height.

Our first attempt to determine H from a number of satellites was in 1962. We obtained 44 values of H from seventeen satellites,[15] leading to the diagram reproduced as Fig. 4.7 The points fell into three separate groups according to their dates, 1958, 1959 and 1960–61 being distinguished. At the time this work seemed difficult, and difficult to verify: we were not fully confident either that the orbits were accurate enough or that all the corrections had been correctly made. In retrospect, the values have proved more reliable than expected: the values of H from the COSPAR International Reference Atmosphere 1972, for the temperatures appropriate in each of the years, have been drawn in on Fig. 4.7. The values derived in 1962 fall within the appropriate areas.

This work pointed to the need for accurate and reliable orbits. There were no more multi-satellite graphs like Fig. 4.7. Instead, in the 1970s, values of H were derived by analysing the orbits of particular satellites specially determined from observations.

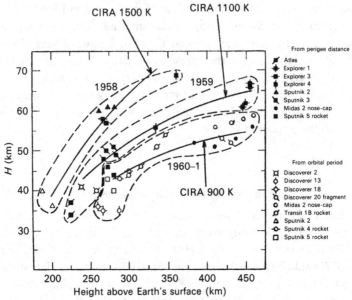

Fig. 4.7. Values of density scale height H obtained in 1962 from the changes in perigee distance and orbital period of various satellites,[15] with later values from *CIRA 1972* for comparison.

How fast does the upper atmosphere go round?

Analysis of the changes in orbital inclination for Sputnik 2 and Sputnik 3 rocket had shown that the actual decrease in orbital inclination was greater than the change theoretically calculated on the assumption that the upper atmosphere was rotating at the same rate as the Earth itself. Thus we concluded (in 1959) that the upper atmosphere, or perhaps just that part of it sampled by these two satellites, was rotating faster than the Earth. Introducing the symbol Λ, Greek capital lambda, for the atmospheric rotation rate relative to the Earth's rotation rate (i.e. the rotation rate in revolutions per day), we were finding values of Λ greater than 1.0. Numerically the value of Λ from the two Sputniks was near 1.25, but the possible error was thought to be at least 0.1 and perhaps greater. So I cautiously concluded in 1961 that the idea of west-to-east winds – or super-rotation as it was often known – 'must at present be treated as rather conjectural'.[16]

Better values of Λ demanded more accurate orbits for high-drag satellites: but there were no new orbits to rival in accuracy those of the early Sputniks. In 1964 the COSPAR meeting was again held in Florence. This time I travelled by air to Pisa, where Byron and Shelley had lived on the Lung' Arno, and then by train to Florence. For the COSPAR meeting I tried to derive better values of Λ from various new satellites, but the harvest was meagre: 'The values of orbital inclination available for these studies are sadly lacking in accuracy',[17] I commented in the paper. But the results were still significant, because the sixteen values of Λ obtained were all greater than 1.0; they ranged between 1.1 and 1.9; their mean was $\Lambda = 1.37$ with a scatter (r.m.s.) of 0.25. (As it happened, all the very high values of Λ were of poor accuracy, and with hindsight it would have been better to have quoted a weighted mean.) The most important new result was from the RAE orbit of Sputnik 3, which gave Λ as 1.1 ± 0.2, to go with 1.2 ± 0.2 from Sputnik 2 and 1.4 ± 0.2 from Sputnik 3 rocket. If the unweighted mean value of 1.37 ± 0.25 for Λ is accepted, it corresponds to a mean west-to-east wind of about 150 ± 100 m/s at a mean latitude of 30° for heights of 190–300 km.

In the absence of more accurate orbits for satellites of high drag, this unsatisfactory evaluation of Λ remained current until 1966, when Diana Scott collaborated with me in a fresh look at this stubborn problem. We first developed a new theoretical equation taking account of terms in e^2 and ce, where e is the eccentricity and c the atmospheric oblateness parameter, which is usually less than 0.2. We found ten new satellites for which the available orbits were not too inaccurate: the average value of Λ that emerged, for heights of 200–300 km, was $\Lambda = 1.27 \pm 0.18$, that is, an average west-to-east wind of 100 ± 70 m/s at latitude 30°. There was also some indication that Λ increased with height.[18]

Soon after this we were able to analyse the orbit of a small satellite called ORS 2, which decayed very rapidly and yielded an accurate rotation rate over a short time interval. The value was $\Lambda = 1.5 \pm 0.1$ for June–August 1966, when the perigee was near the sunset line at a height of 200 km. Fig. 4.8 shows the observational values of inclination and the theoretical curve for $\Lambda = 1.5$, which fits well. This value of Λ, corresponding to a 200 m/s west-to-east wind, 'suggests that there is a particularly strong west-to-east wind at about sunset,' we concluded in a note in *Nature*.[19]

Several other useful orbits emerged in 1967, and we were able to derive nine new longer-term values, including a much lower value for ORS 2 over its whole life. These nine values showed an increase in Λ with height, from about 1.1 at 200 km to about 1.3 at 250 km.[20]

The web was beginning to unravel. The rotation rate seemed to depend on the perigee height and the local time at perigee: it increased with height, and the strongest west-to-east winds were in the evening. Would the rotation rate also depend on latitude, season or solar activity?

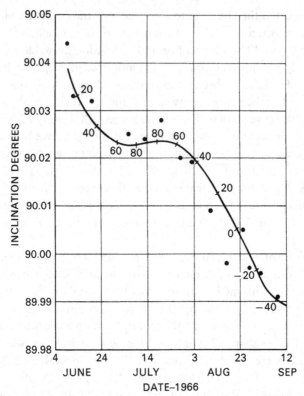

Fig. 4.8. Values of orbital inclination for ORS 2 from NASA Bulletins, with theoretical curve for $\Lambda = 1.5$. (Numbers on the curve indicate the perigee latitude in degrees.)

In 1968 we searched for more satellites suitable for evaluating atmospheric rotation rates. Eleven useful new orbits were found and, with twenty-one previous values that seemed reliable, led to a much better diagram for the variation of Λ with height, that showed an increase 'from about 1.1 at 200 km height to about 1.35 at 300 km,'[21] that is, a west-to-east wind increasing from 40 m/s at 200 km height to 140 m/s at 300 km, on taking the usual average latitude of 30°. Above 300 km the results were rather scattered. No evidence was found of variations from year to year, or with solar activity.

That is how things stood at the end of 1969, but we were puzzled by one result, from Transit 1B over its seven-year life from 1960 to 1967. The inclination had fallen rapidly during 1962: if atmospheric rotation was the cause, a ridiculously large value of Λ would be needed (about 10); the decrease must be caused by another perturbation as yet unknown, but there was no sign of any sudden jump that might point to collision with a meteorite. 'No satisfactory explanation can be given', we remarked.

Theoretical models of upper-atmosphere winds devised by physicists in the 1960s did not support our finding that the winds were of order 100 m/s. Their models gave much weaker winds, and they looked askance at us. Towards the end of the 1960s, more realistic models began to appear, which were in better agreement with our observational values. The results obtained by measuring vapour trails released from rockets also encouraged us by showing that winds were generally from east to west in the morning and from west to east in the evening, sometimes with speeds in excess of 100 m/s.

Hard at work on orbital theory

I have run through these expanses of upper-atmosphere research as though all the techniques were stored ready-made in a cupboard, and only had to be taken out and applied. That was far from the truth, and much intellectual effort went into the development of theory, which was needed not only in the atmospheric researches but also in devising the computer programs for orbit determination and in evaluating the gravity field.

In the theory for the effect of air drag on orbits, we were striving to eliminate the most damaging over-simplifications. A spherical atmosphere was assumed in Part I, an oblate atmosphere in Part II; in both parts the upper limit for eccentricity was 0.2, but satellites with higher eccentricities were becoming more common.

In Part III, published in 1962, I extended the theory[22] to orbits of high eccentricity, between 0.2 and 1, assuming a spherical atmosphere because Part II showed that atmospheric oblateness did not have much effect on high-eccentricity orbits. For high eccentricity the mathematical techniques were different, but the results came out quite cleanly, linking well with the small-e theory. The equation for da/dx quoted on p. 62 for small e, looks different for large e, being

$$\frac{da}{dx} = 1 + \frac{1-e}{2z(1+e)} + O\left(\frac{1}{8z^2}\right),$$

where $z \,(= ae/H)$ is now always greater than 20. On integration, however, the equation for perigee distance is somewhat similar to that on p. 62, namely

$$r_{p0} - r_p = \tfrac{1}{2}H \ln \frac{e_0(1+e)}{e(1+e_0)} + O(0.005\,H).$$

The theory also gave the variation of eccentricity with time: the variation of e^2 with time is not linear, as it is for small e, but follows a curve that has a gentle slope initially and becomes steeper towards the end of the life. The variation of z with time is unexpectedly simple, however: it decreases almost linearly as e decreases from e_0 to 0.2. The variation of orbital period with time is rather peculiar for high eccentricities: when $e < 0.2$ the slope of the curve steadily steepens; but if the initial eccentricity is high, the slope becomes more gentle as time goes on, until e decreases to $\frac{1}{3}$, when the slope steepens.

An unforeseen bonus from the high-eccentricity theory was a better version of the standard equation for air density in terms of $\dot{T}$. The low-eccentricity theory had given the term in curly brackets in the equation on p. 54 as

$$\left\{1 - 2e + \frac{5e^2}{2} + O\left(e^3, \frac{1}{8z}\right)\right\};$$

the high-eccentricity theory showed that this term could be written in a more accurate and more general form as

$$\left\{(1-e)^{\frac{1}{2}}(1+e)^{-\frac{3}{2}} + O\left(\frac{1}{8z}\right)\right\},$$

for any e greater than about 0.02.

In Parts I–III the scale height H was assumed constant, so that the variation of density ρ with height y was exponential, with ρ varying as $\exp(-y/H)$. By now we knew that H tended to increase with height, and in Part IV Graham Cook and I redeveloped the theory for an atmosphere in which H varied linearly with height y. We assumed that

$$H = H_p + \mu(y - y_p)$$

at heights above the perigee height y_p and that $\mu < 0.2$, which in general seemed to be true. In this theory,[23] published in 1963, our hope was to show that the constant-H equations could still be used if H was evaluated at a particular height, and this hope was nearly always realized. For example, we found that the equation for the change in perigee height given on p. 62 still applies if H is evaluated at a height $\frac{3}{2}H_{p0}$ above the initial perigee height y_{p0}. In practice there were now graphs available which showed the variation of H with height (e.g. Fig. 4.7), so it was just as easy to read off the value of H at this new height as at the perigee height. This discovery also showed that the values of H determined by analysing the decrease in perigee height applied at a height $\frac{3}{2}H$ above the perigee – a result used in arriving at Fig. 4.7.

The variation of eccentricity with time is directly affected by μ: the necessary extra term is usually small, but can alter e/e_0 by 0.02 if $\mu = 0.2$ and $t/L = 0.9$. The equation for determining air density from $\dot{T}$ needs only slight alteration when the assumption of constant H is abandoned. If $\mu = 0.1$ is taken as a better value than the previous $\mu = 0$, the equation on p. 54 needs to have the constant 0.158 changed to 0.157 – a change smaller than the likely errors from other sources.

These details of the orbital theory may seem dull and difficult, but this was the nature of the beast. We worked hard on the theory because it was the basis for the research: but few people understood it, so we didn't expect appreciation. Sometimes the papers were presented at conferences, but usually with little more impact than a bag of feathers. Most of the conferences were rather dull too, but I remember one in May 1962 at Paris, where I was one of the organizers and gave a talk about the results of Part IV. An evening reception by the Mayor of Paris unfortunately coincided with an exceptional downpour of thundery rain. I eventually arrived very late, looking like a drowned rat. Even so, I was brought before the Mayor and, to my astonishment, he presented me with the Silver Medal of the City of Paris. I never knew why – that's life! – but perhaps it was in appreciation of the orbital theory: I hope so.

Later in 1962 I was asked by the publishers Butterworths to write a mathematical book setting out our theories of the effects of air drag on orbits. I didn't like to refuse, for this was appreciation indeed, and I worked on the book throughout 1963. It included the theory up to Part IV and was published in 1964 under the title *Theory of Satellite Orbits in an Atmosphere*. My experience with this book was rather traumatic. The first set of proofs had more than a thousand errors – not surprising for such a heavily mathematical book. After a fierce battle I succeeded in having a second set of proofs, which still had about 200 errors: but I was not allowed to see a third set. In the published book most of the 200 errors had been corrected, but a new crop had appeared. some of these were quite weird: on p. 3, for example, 'Ω' became 'O' and 'eastwards' became 'eastwares'. I vowed never again to write a mathematical book, and kept that vow until the mid-1980s, when I was persuaded to enlarge and update the 1964 book. The Butterworth book was in a series, but I retained the copyright myself and was assured I would be told if the book was remaindered. In fact the whole series was pulped without the authors' knowledge a few years later: when I protested, the publishers apologized and admitted they were open to legal action. But that was pointless, as it would not resurrect the book. In the 1970s I grew accustomed to seeing photocopied copies being used in universities.

To return from these horror-stories to the orbital theory, the next step was to try to tame the day-to-night variation in air density. The first problem was to devise a realistic yet fairly simple model for the variation in density. We took the density ρ at a distance r from the Earth's centre to be dependent only on the angular distance ϕ from the point B, the 'centre of the diurnal bulge', where the density reached its daytime maximum: the variation with ϕ was taken to be sinusoidal. With an exponential variation of density with height, ρ was expressed as

$$\rho = \rho_0(1 + F\cos\phi)\exp\left\{-\frac{(r-r_0)}{H}\right\},$$

where ρ_0 is the density at distance r_0 from the Earth's centre at $\phi = 90°$. Despite its simplicity, this model usually gave values of density close to those actually experienced by a satellite as it moved round its orbit, if F and H were well chosen. This was the basis for Part V of the theory,[24] published in 1965. The work proved quite difficult, but eventually we derived expressions for all the usual parameters – perigee distance, eccentricity and so on. The most peculiar results occur when the orbit is nearly circular: if the value of ϕ at perigee, ϕ_p, varies only slightly and can be taken as constant (as might happen if the lifetime in Phase 2 is only a few days), the normal linear decrease of e^2 with time (down to zero) does not apply. Instead the quantity $(z + F\cos\phi_p)^2$ usually decreases linearly to zero. As a further twist in the story, the very idea of taking ϕ_p constant may be untenable, because the asymmetrical drag produced by the day-to-night variation in density can greatly alter the angular position of the perigee. These subtleties of near-circular orbits were postponed for a further paper, Part VI.

Before attempting Part VI, however, we needed a better method for displaying the full effects of gravitational forces on near-circular orbits *in vacuo* – a picture to which the effects of drag could be added. Graham Cook wrote an important paper[25] clarifying the effects of zonal harmonics on near-circular orbits *in vacuo*. Instead of working with eccentricity e and argument of perigee ω, he used the variables ξ and η, defined by

$$\xi = e\cos\omega \qquad \eta = e\sin\omega.$$

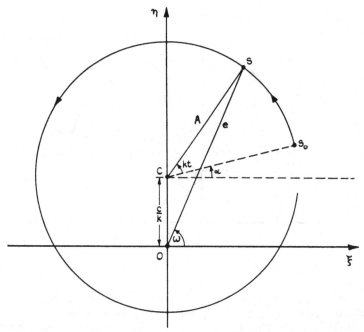

Fig. 4.9. Representation of an orbit *in vacuo* in the (ξ, η) plane – S moves on a repeating circle of radius A centred at the point $(0, c/k)$, with constant angular velocity k.

He showed that the usual steady change in ω caused by the second and other even harmonics in the gravity field was modified by the effects of the odd zonal harmonics (third, fifth, etc.) in such a way that the values of ξ and η trace out a circle in the (ξ, η) plane, as shown in Fig. 4.9. The centre C of the circle is at a point on the η-axis at a distance c/k from the origin, where c depends on odd zonal harmonics $(J_3, J_5, \ldots)$ and k depends on J_2.† For an orbit *in vacuo*, the radius A of the circle remains constant and rotates at a constant rate. So the circle is repeated indefinitely. In practice c/k usually has a value near 0.001, and if A exceeds c/k, as in Fig. 4.9, the eccentricity oscillates between $A + c/k$ and $A - c/k$, with ω going through a complete cycle of 360°, though not at a constant rate. For orbits of very small eccentricity, the circle in Fig. 4.9 may be entirely above the ξ axis: e then oscillates between $c/k + A$ and $c/k - A$, and ω does not go through a full cycle but remains between 0° and 180°, and can sometimes be confined

† Explicitly, $\dfrac{c}{k} = \dfrac{R \sin i}{2 a J_2} \{ -J_3 + \text{terms in } J_5, J_7, \ldots \}$.

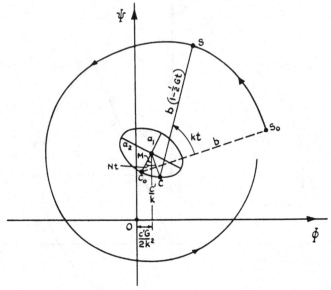

Fig. 4.10. In the presence of drag,[26] S moves in the (Φ, Ψ) plane with angular velocity k on a circle of decreasing radius $b(1 - \frac{1}{2}Gt)$, whose centre C moves round a small ellipse with angular velocity N. Note that $c' = ac/H$, $\Phi = a\xi/H$ and $\Psi = a\eta/H$ (to link with Fig. 4.9).

to a small sector near $\omega = 90°$. This representation of near-circular orbits was most important for determining values of odd zonal harmonics, as will be seen later.

Armed with this pictorial device, Graham Cook and I examined the combined effects of air drag and gravity on near-circular orbits, to produce Part VI of the theory.[26] Instead of working with z and ω, we used

$$\Phi = z \cos \omega \quad \text{and} \quad \Psi = z \sin \omega.$$

We found that, when the drag is not too large, the behaviour of the orbit could be portrayed in the (Φ, Ψ) plane as a circle of slowly decreasing radius. The centre C of the circle, instead of being fixed, moves on a very small ellipse centred at a point just off the Ψ axis. Fig. 4.10 shows the choreography of this complex ballet: G is a constant dependent on the drag, $1/G$ being approximately the lifetime. The diagram looks complicated but has been very useful for removing the effects of drag from near-circular orbits when evaluating the effects of the odd zonal harmonics, i.e. evaluating c/k.

As compensation for the brain-teasing effort of trying to figure out the complex gyrations of Fig. 4.10, I ought now to produce some racy personal adventures. Alas, I have none to offer: the orbital theory was ivory-tower work. However, the papers on orbital theory, published in the Royal Society's journals, probably contributed to my election as a Fellow of the Royal Society in March 1966: who can tell, now that most of those concerned are dead? I was the first FRS to be elected for contributions to space science, and it was the signal that space had become respectable: a second space scientist, Robert Boyd, was elected three years later, and others followed.

I have much enjoyed my twenty-five years as a Fellow of the Royal Society: the Society has been the best organization I have encountered. Initially, however, the Society must have thought me an *enfant terrible* – I was 38, relatively infantile by Royal Society standards (the average age of Fellows was over 60). One of my first suggestions as a Fellow was that the Society abandon compulsory evening dress for its Soirées. In my view this gave the Society a bad image, making it seem elitist and antiquated. My plea was ignored, so I never went to a Soirée until 1987, when I was an exhibitor for the tercentenary celebrations of Newton's *Principia*. Needless to say, I did not wear a penguin suit, but no one dared to object (probably because I was by then chairman of the National Committee for the History of Science). And the Society did scrap evening dress for its anniversary receptions in 1988. These sartorial pinpricks are trivia, but may have been helpful in keeping me a little independent of the Society's conventions.

The most pleasant aspect of my years as a Fellow has been my affinity with the Royal Society's Library, and this attraction began in the early 1960s. In 1967 Professor Rupert Hall and I, as members of the Library Committee, suggested that the Library should cease to try to keep up with all science (as it then did) and should become primarily a history-of-science library. This radical plan was adopted – successfully, as it turned out. The Royal Society's Library is the most beautiful I know, with ceiling paintings and much gold leaf; the staff have always been kind and most efficient; and the book-stock is well adapted to my interest in the history of science. No wonder the Library has retained my affection.

After this digression it is time to return to Part VI of the orbital theory. The gyrations of Fig. 4.10 apply only when the air drag is relatively slight. What happens at the other extreme, in the last few days of the satellite's life, when drag is overwhelmingly severe? Part VI provided the answer, by showing that, as decay approached, the various parameters tend towards specific 'final' values, given the suffix f. The value of ϕ at perigee, ϕ_p, becomes greater than 90° as decay approaches. The perigee moves towards the point on the orbit which is closest to the minimum-density point: if the orbit happens to go through the minimum-density point, $\phi_{pf} = 180°$; if not (as is usual) ϕ_{pf} is between 90° and 180°. And the final value of z tends towards $-F\cos\phi_{pf}$. If an anthropomorphic analogy helps, you can think of the orbit as wriggling its perigee towards the point of minimum density on the orbit, and simultaneously adjusting its eccentricity so as to equalize the drag round the orbit as far as possible. It is trying to imitate a circular orbit in a spherical atmosphere, and also postponing its death sentence for as long as possible. These results proved useful in interpreting the behaviour of orbits in their last few days of life, as later chapters will show.

Parts I–VI dealt with the drag acting in the orbital plane: this is the major part of the drag, but the smaller out-of-plane forces also need to be considered. The out-of-plane forces arise because of the rotation of the atmosphere and produce small changes in inclination, as discussed in the previous section. The theory for the effect of atmospheric rotation derived in 1960 was in need of extension: in 1966 extra terms were added to give better accuracy, as mentioned on p. 97. In 1969 the theory was redeveloped for an atmosphere in which the density scale height H varied linearly with height, with $\mu\,(= \mathrm{d}H/\mathrm{d}y)$ taken as less than 0.2, as in Part IV. We found[27] that the effects of μ could be reduced to a negligible level by evaluating H at a height $\frac{3}{4}H$ above the perigee, whatever the eccentricity. (This may seem odd for near-circular orbits on which the satellite never reaches such a height: but H has little effect on such orbits, so there is no need to modify the rule.) The theory also showed that, for orbits of eccentricity greater than about 0.02, the average height at which the inclination-changing forces act can best be taken as $\frac{3}{4}H$ above the perigee. Previously we had guessed that both these heights would be $\frac{1}{2}H$ above the perigee. The change to $\frac{3}{4}H$ altered the previous results for the atmospheric rotation rate Λ by less than 0.7%, which was not significant.

Atmospheric rotation is a measure of west–east winds, but what about north–south (meridional) winds? I examined their effects on inclination, for orbits of eccentricity less than 0.2, in a paper[28] published in 1966, which showed that the effects of a consistent meridional wind cancelled out over every half-cycle of the angular travel of the perigee, if the orbit did not contract much. For long-lived satellites, therefore, the effect of a constant meridional wind would generally be negligible, and particularly so for near-polar orbits, for which the meridional wind is nearly in-plane. However, meridional winds could have significant effects for special orbits, for example if the perigee position moved little, or if the meridional wind happened to vary with the same periodicity as the perigee.

Nearly all the orbital theory so far mentioned has involved air drag, but theory for all the other main effects on orbits was also being developed at the RAE by Robin Merson, Graham Cook, Bob Gooding and also by Russell Allan, who arrived in the early 1960s and wrote several seminal papers in orbital theory. One was on the effects of radiation pressure from the Sun and the gravitational attractions of the Sun and Moon;[29] another of Allan's papers explained how the variation of gravity with longitude would affect geostationary satellites.[30] After that, he turned his attention to the effects of resonances with the gravity field, leading on to the main stream of our work in the 1970s. Graham Cook wrote an important paper[31] specifying the effects of lunisolar perturbations explicitly (it remains a 'bible' for analysts today), as well as pioneering work on the effects of tesseral (longitude-dependent) harmonics and of aerodynamic lift.[32] This last paper was important because it showed that the effects of lift – i.e. forces perpendicular to the direction of motion – should be negligible except for flat-plate satellites. Graham Cook and Diana Scott also developed a computer program called PROD for evaluating zonal harmonic, lunar and solar gravitational effects on drag-free orbits: the program was extensively used in the 1970s and 1980s to determine the lifetimes of high-eccentricity satellites and to remove the basic gravitational and lunisolar perturbations from orbits due to be analysed for other purposes. Robin Merson developed the orbital theory underlying the PROP orbit refinement program, and, as mentioned earlier in this chapter, Bob Gooding produced a comprehensive orbital theory covering all kinds of gravitational effects in one scheme, as well as writing the orbit determination programs. There was also work on spacecraft propulsion. Harry Hiller wrote three definitive papers on the best mode of transfer between non-coplanar elliptic orbits,[33] I studied the enlargement of orbits by microthrust – a 'reverse' of drag – and George Burt, the Head of Space Department, investigated the optimum procedures for low-thrust electrical propulsion.[34] To say more on all this work would take me too far from the main theme of orbit analysis, and I now return to the geophysical researches made possible by the theory.

The flattening of the Earth: even zonal harmonics

During the 1960s we were evaluating the even zonal harmonics of the Earth's gravitational field J_2, J_4, J_6, ... from the rate of rotation of the orbital plane, and the odd zonal harmonics J_3, J_5, J_7, ... from the variations in eccentricity. This section covers the even harmonics, which are symmetrical about the equator; the next section is devoted to the odd harmonics.

Our best evaluation of the even harmonics before 1961 was for J_2, J_4 and J_6, as given in Table 3.1 and repeated in the first line of Table 4.1: but these results relied on only three orbits, with no redundancy. In 1963, working with Graham Cook, I found several further suitable satellites, and we made a new determination of the J coefficients. But this was almost immediately overtaken by events, as a number of new and more accurate orbits became available. Consequently another new evaluation was made in 1964, from seven satellites of low drag. As these had a good spread in orbital inclination, from 28° to 96°, the new values seemed likely to be reliable, and were quite independent of the previous set. Our preferred values, published in 1965, were for only four coefficients, $J_2 - J_8$, though a solution for six coefficients was also given.

Table 4.1 lists the 1960 values, a 1964 solution by Kozai, our preferred values of 1965 (with s.d. doubled), and values obtained in 1965 by David Smith from close satellites and by Alan Cook from distant satellites. Values from GEM T3 are included for comparison, with s.d. doubled to give an impression of the likely maximum error.

The results in Table 4.1 speak for themselves. The first three evaluations of J_2 in 1964–65 (all independent) not only agree with each other but also agree to within 1 part in 50000 with the 1992 value from GEM T3. Even though it may imply self-praise, I think this was an extraordinary achievement so soon after the satellite launchings. The pre-satellite error in J_2, expressed as a fraction of total gravity, was 8000×10^{-9} (see Table 3.1); the error in the 1964–65 values was, as it now appears, 14×10^{-9}; and the 1992 nominal s.d. (doubled) is 1×10^{-9}. Table 4.1 also shows that reliable values of J_4 and J_6 were available by 1965, though the same could not be said for J_8 (our alternative solution gave $10^6 J_8$ as 0.13, instead of 0.44). Comparison with Table 3.1 on p. 66 shows that Kozai's new value of J_2 differs by 15 s.d. from his anomalous 1961 value, and his 1964 values, though still with too low an s.d., have proved to be marginally the best of the three sets: this is not surprising, as he was working at the Smithsonian Astrophysical Observatory, where the most accurate orbits of low-drag satellites were being determined.

The differences between the three sets of values obtained in 1964–65 were only significant for satellites with inclinations lower than 20°, where no good data existed. We thought it was not worth attempting further evaluations of even harmonics until there were some good orbits for near-

Table 4.1. *Values of even zonal harmonic coefficients* J_2, J_4, J_6, J_8

Author, and year of publication	$10^6 J_2$	$10^6 J_4$	$10^6 J_6$	$10^6 J_8$	Published in:
King-Hele, 1960	1082.79 ± 0.15	−1.4 ± 0.2	0.9 ± 0.8	—	*Nature*, **187**, 490–491
Kozai, 1964	1082.64 ± 0.01	−1.65 ± 0.02	0.65 ± 0.03	−0.27 ± 0.05	*Publ. Ast. Soc. Japan*, **16**, 263–284
King-Hele & G. E. Cook, 1965	1082.64 ± 0.04	−1.52 ± 0.06	0.57 ± 0.14	0.44 ± 0.22	*Geophys. J. Roy. Astronom. Soc.*, **10**, 17
Smith, 1965	1082.64 ± 0.08	−1.70 ± 0.25	0.73 ± 0.40	−0.46 ± 0.42	*Planet. Space Sci.*, **13**, 1151
A. H. Cook, 1965	1082.65 ± 0.1	−1.61 ± 0.15	0.73 ± 0.40	—	*Geophys. J. Roy. Astronom. Soc.*, **10**, 181
GEM T3, 1992	1082.626 ± 0.001	−1.619 ± 0.003	0.539 ± 0.004	−0.20 ± 0.01	NASA Tech. Memo. 104555

equatorial satellites. We envisaged a delay of a few years, but in fact we made no further evaluations of even zonal harmonics. This was not a deliberate decision, but a result of different interests, and fewer staff, in the 1970s.

In the early 1960s many people were interested to hear that we had managed to change perceptions about the shape of the Earth. Our 1965 value of J_2 indicated that the difference between the equatorial and polar diameters of the Earth was 170 m less than had been thought in pre-satellite days: it was 42.77 km rather than 42.94 km. Expressed as a fraction of the equatorial diameter, this corresponded to a flattening f of 1 part in 298.26. Our deflattening of the Earth, or making it more nearly spherical, took the fancy of the listening public and triggered many requests for radio talks as well as 'live' lectures in the early 1960s. In 1963 I was invited to give the third Duke of Edinburgh's Lecture of the Royal Institute of Navigation. My subject was 'The Shape of the Earth', and about half of the talk revolved round past ideas on the subject. I found little to guide me, on the eighteenth-century work in particular, and I had to go back to the original books of Maupertuis, Cassini and others (the ones I saw at the Library of Congress in August). This research proved rather fascinating and, together with my short biography of Erasmus Darwin that Macmillan published in October, was edging me towards the history of science. The lecture itself, with the Duke of Edinburgh in the chair, was well received, though most people found the historical stories easier to swallow than the harmonics.

The Earth's pear-shaped profile: odd zonal harmonics

As explained in Chapter 3, the lack of symmetry between the northern and southern hemispheres of the Earth, associated with the odd harmonics J_3, J_5, J_7, ..., causes an asymmetry in the gravity field, which produces an oscillation in the perigee distance r_p: this is equivalent to an oscillation in the eccentricity e, because $r_p = a(1-e)$ and a remains constant in the absence of air drag. By measuring the magnitude of this oscillation, as exhibited by Vanguard 1, John O'Keefe and his colleagues obtained a value for J_3 (-2.4×10^{-6}), published in 1959. Although the higher odd harmonics (J_5, J_7, ...) were ignored, this 1959 value has proved surprisingly good. In the ensuing five years, several further determinations of J_3 and some higher odd harmonics were made: Robert Newton and others (1961) worked with Transit satellites; Yoshihide Kozai (1961) used SAO orbits;

the first UK evaluation was by David Smith (1963). In 1964 Kozai obtained values much better than his previous set: indeed they are within about 1 s.d. of the 1992 values. All we had done so far was to watch and wait – and give priority to orbital theory.

Independent evaluations at the RAE began in 1964: Graham Cook and Diana Scott collaborated with me in this work, which was quite extensive. The method was basically very simple, however. The sinusoidal oscillation in eccentricity produced by the odd zonal harmonics is given by the equation

$$e = e_0 + (R/2aJ_2) \sin i \sin \omega \{-J_3 + \text{terms in } J_5, J_7, \ldots\},$$

where e_0 is here the value of e when $\omega = 0$, and air drag is assumed to be negligible. As the numerical value of J_3 is negative, the equation shows that usually the eccentricity rises to a maximum when the perigee is at the northern apex ($\omega = 90°$) and sinks to a minimum when the perigee is at the southern apex ($\omega = 270°$).

Thus all we had to do was to find an accurate orbit over a cycle of ω (usually several months); to remove the effects of other forces such as drag and lunisolar perturbations; and to plot out the values of eccentricity, to produce (with luck) a perfect sine wave. The amplitude of this oscillation gives a numerical value, K say, so that

$$(R/2aJ_2) \sin i \{-J_3 + A_5 J_5 + A_7 J_7 + \ldots\} = K,$$

where $A_5, A_7, \ldots$ are easily-calculated numerical values which depend on the satellite's inclination i and semi major axis a. In practice the sine wave is never quite perfect, but often nearly so. Fig. 4.11 shows the observational values for Vanguard 2, the values after removal of perturbations, and the fitted sine curve: the resulting value of K was nominally accurate to better than 1%, namely $(444 \pm 3) \times 10^{-6}$. (The curve shown in Fig. 4.11 is from a later paper (1969), but the fitting was very similar in 1964.)

Such values of K for, say, six satellites at different inclinations would give six numerical equations for $J_3, J_5, J_7, \ldots$, and these could be solved for up to six of the J coefficients, though caution would suggest evaluating fewer than six. I have used six as an example, because that was the number of satellites finally accepted in our first evaluation,[35] published in 1965. Their inclinations were 29°, 33°, 47°, 52°, 67° and 96°: four of the orbits were determined by the SAO, one by Bob Gooding (Ariel 2) and one from Doppler tracking (Transit 4A). We evaluated four of the J coefficients, the

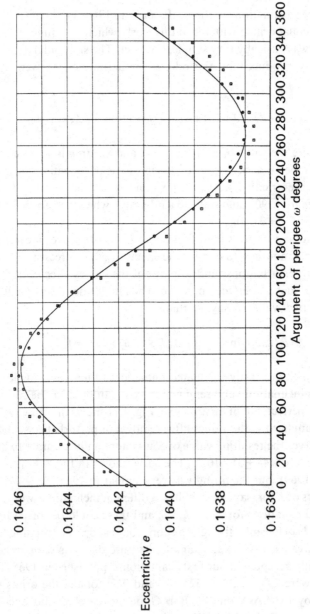

Fig. 4.11. Variation of eccentricity with argument of perigee for Vanguard 2, with fitted sine curve.[37] The squares show the observational values; the circles show the values cleared of lunisolar perturbations.

'1965' set in Table 4.2. On this our first attempt we cautiously assigned 'realistic' errors, which in retrospect have proved to be about twice as large as they need have been.

Cheered by the success of this first foray into the field, we soon produced a new set of values,[36] published in 1967. This time we aimed for the widest possible range of inclination and semi major axis, and picked seventeen satellites, including the previous six. Three of the seventeen were omitted, and fourteen were used in the preferred final solution, for seven coefficients, the first four of which appear in Table 4.2. This 1967 solution has much smaller errors than its predecessor, and the values of J_3–J_7 have stood the test of time well, being within 1 s.d. of the GEM T3 values, which in Table 4.2 are again given with the nominal s.d. doubled.

This further success induced us to continue the work, and in 1968 we added eight new orbits, to improve the coverage in inclination, though there were still no suitable satellites at inclinations less than 28°. The results of this analysis of 22 satellite orbits were published[37] in 1969. We were keen to extend the evaluation to high degree, and we found it was possible without loss in accuracy to set J_9, J_{11}, J_{13} (and some higher Js) equal to zero. If this rather bold step was taken, some higher harmonics, such as J_{15}, J_{21} and J_{27}, could be evaluated. This is the set (1) given in Table 4.2, and it has very low s.d.: as the zeros make it less satisfactory for comparison, the best solution without zeros obtained in the 1969 paper is also given, as set (2). Though nominally less accurate, this second set of values is within half a standard deviation of GEM T3 for all four coefficients (and also for J_{11}). The set (1) is also good for J_3, J_5 and J_7, though the errors are slightly optimistic.

Kozai made a new determination of odd zonal harmonics in 1968, also given in Table 4.2. His values, though nominally more accurate, are not so close to GEM T3 as either ours or his own 1964 values. Though we were late starters, we caught up with the rest, it now seems. Such retrospective judgements were not possible at the time; instead, people looked to see whether the independent evaluations by Kozai and by ourselves were consistent. Fortunately both evaluations were good – and consequently consistent.

Our methods in 1967 and 1969 were similar to those for 1965, already described, with one exception. In 1965 we had evaluated K only by fitting sine curves, as in Fig. 4.11. For nearly-circular orbits, however, the variation is not exactly sinusoidal, and the fitting is much more efficient and accurate if use is made of the representation of the orbit as a circle in the (ξ, η) plane, as in Fig. 4.9. This technique was used in the 1967 and 1969

Table 4.2. *Some values of odd zonal harmonic coefficients, 1959–1969*

Author, and year of publication	$10^6 J_3$	$10^6 J_5$	$10^6 J_7$	$10^6 J_9$	Published in:
O'Keefe, Eckels & Squires, 1959	-2.4 ± 0.3	-0.1 ± 0.1	—	—	*Astronom. J.*, **64**, 245–253
Newton, Hopfield & Kline, 1961	-2.36 ± 0.14	-0.19 ± 0.10	-0.28 ± 0.11	—	*Nature*, **190**, 617–618
Kozai, 1961	-2.29 ± 0.02	-0.23 ± 0.02	—	—	*Astronom. J.*, **66**, 8–10
Smith, 1963	-2.44 ± 0.07	-0.18 ± 0.03	-0.30 ± 0.03	—	*Planet. Space Sci.*, **11**, 789–795
Kozai, 1964	-2.55 ± 0.02	-0.21 ± 0.02	-0.33 ± 0.04	-0.05 ± 0.06	*Publ. Ast. Soc. Japan*, **16**, 263–284
King-Hele, Cook & Scott, 1965	-2.56 ± 0.10	-0.15 ± 0.15	-0.44 ± 0.2	0.12 ± 0.2	*Planet. Space Sci.*, **13**, 1213–1232
King-Hele, Cook & Scott, 1967	-2.53 ± 0.02	-0.22 ± 0.04	-0.41 ± 0.06	0.09 ± 0.06	*Planet. Space Sci.*, **15**, 741–769
Kozai, 1968	-2.56 ± 0.01	-0.17 ± 0.01	-0.42 ± 0.02	-0.02 ± 0.03	SAO Spec. Rpt, 264
King-Hele, Cook & Scott, 1969 (1)	-2.54 ± 0.01	-0.21 ± 0.01	-0.40 ± 0.02	(0)	*Planet. Space Sci.*, **17**, 629–644
(2)	-2.53 ± 0.03	-0.25 ± 0.05	-0.31 ± 0.09	-0.18 ± 0.13	
GEM T3, 1992	-2.5325 ± 0.0004	-0.227 ± 0.001	-0.354 ± 0.004	-0.12 ± 0.01	NASA Tech. Memo. 104555

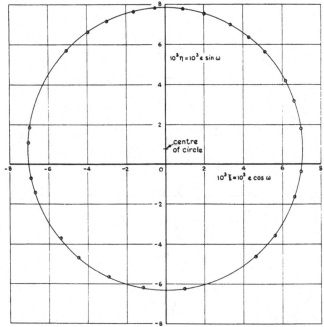

Fig. 4.12. Values of ξ and η for Anna 1B, with fitted circle.[37]

solutions, and an example of fitting the observational values with a circle is shown in Fig. 4.12 for the satellite Anna 1B at inclination 50°. The mean eccentricity is 0.007, and the value of K is given by the distance between the origin and the centre of the circle, namely 0.771×10^{-3}.

In the 1969 paper we gave a picture showing the profile of the sea-level surface of the Earth (the *geoid*) implied by our 1965 even zonal harmonics and set (1) of the 1969 odd zonal harmonics, as given in Table 4.2. This picture is reproduced as Fig. 4.13; a later version appears in Chapter 5. With these values of the odd harmonics, sea level at the north pole is about 41 m further from the equator than is sea level at the south pole. The profile of Fig. 4.13 is an average meridional section of the Earth sliced through the poles: over most of the latitude range the profile should be accurate to about 1 m, though the likely error becomes much greater close to the poles. Many people are surprised that orbit analysis, with observations accurate to perhaps 100 m, can achieve such good accuracy. The reason is easy to see: the 40 m deformation in the Earth's shape leads to an oscillation in satellite perigee height of 5–10 km amplitude (the exact value depends on the inclination). So, if you determine the orbit correct to about 100 m, you can measure the amplitude of the oscillation – and hence the 40 m

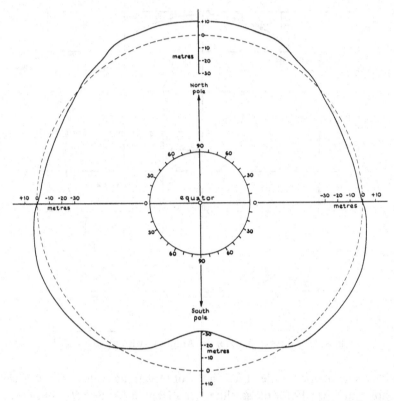

Fig. 4.13. Height of the geoid (solid line) relative to a spheroid of flattening 1/298.25 (broken line), as given by values of odd zonal harmonics[37] published in 1969 (set 1). This is a polar section averaged over all longitudes. The geoid can be regarded as the mean sea-level surface.

deformation – with an accuracy of 1–2%, which corresponds to less than 1 m in geoid height.

In the 1960s the pear-shaped Earth was a sure-fire winner in the publicity game. People liked the idea of a dry subject producing something juicy and homely for them to bite on, and the subject was even considered popular enough for an article in *Scientific American*, published in the October 1967 issue and embellished with coloured diagrams. Scientists in other fields liked the pear-shape image too, and our papers on the odd zonal harmonics attracted more interest (and requests for reprints) than those on air density. Sometimes the results were presented at conferences, of which the most memorable was the second symposium on 'the use of satellites for geodesy', which took me in 1965 to what instantly became my favourite conference centre, Lagonissi in Greece, about 20 miles south of Athens.

This was a large holiday hotel on a peninsular jutting out into the pellucid sea, with outlying bungalows set round rocky bays basking from dawn to dusk in seemingly perpetual sunlight. The conference did have its serious side, with the usual deluge of scientific papers and discussions which led the Ordnance Survey to acquire the Hewitt cameras two years later. But it was something of a holiday, with evening excursions to the Acropolis and Sounion, as well as midday visits to the beaches.

The real world in quite good shape too

In swinging so swiftly through the sixties, I may have given the impression that nothing mattered except chasing satellites and their orbits. Did current events pass me by like a ship in the night? Not altogether: rumblings from the real world were audible in the ivory tower, but the effects were varied and not always correctly interpreted. I did not appreciate at the time my great privilege in having an annual salary greater than the price of a three-bedroomed house, though I was only a research scientist, not a financier or administrator. This statistic is, I think, so arresting that repetition cannot wither it. Now that I have to live on a total ('inflation-proofed'!) pension less than one tenth of the current house price, I regret never having saved money then. Buoyed up by the rising standards of the 1950s, I did not foresee that the real wealth of the UK would decline in the 1970s and 1980s. Nor did I realize that the status of both scientists and civil servants would undergo a startling change from great respect to near-contempt.

Another exhilarating feature of Britain in the 1960s was the efflorescence of the 'caring society', with a strong consensus in favour of a better life for (nearly) everyone. The outlook seemed set fair, with the great expansion of the universities to bolster the idea, and I had no premonition that the 'grab-for-yourself' society of the 1980s was just round the corner.

The bad feature of the 1960s was the international tension and inhumanity. The Cuban missile crisis brought the world close to disaster in 1962, and the fall-out from nuclear testing in the atmosphere was most worrying in the early 1960s. Famine in poor countries seemed likely to continue, or worsen, and few seemed to care.

Of much greater impact than the outside 'real world' was the nearer real world, at home, where two lively young daughters well earned a friend's nickname, 'the mighty atoms', and kept me alert. Apart from this, my main outside influence was from international conferences. An important one that has so far slipped through unmentioned was the COSPAR meeting at Washington in April–May 1962, which included the first symposium on 'the use of satellites for geodesy' (the predecessor of the Lagonissi meeting in 1965). It was in Washington that I first met most of the leading scientists in satellite geodsey, and the discussions influenced our future choice of research topics. The 1962 COSPAR meeting also took the decision to change the international designations of satellites. The first satellite launched in 1963 would be 1963-01, the second 1963-02, and so on; previously the Greek letters had been used, so that Sputnik 1 was 1957α and Sputnik 2 was 1957β.

For me the 1962 Washington meeting had other memorable features. This first flight across the Atlantic was very pleasant because (need I say?) I was entitled to first-class travel, with plenty of leg-room and empty seats around. I arrived ready to dislike Washington, but instead I was captivated, particularly by the Mall, the cleanliness, and the kindness of the people. It has gone downhill since.

It was in 1962 that COSPAR decided to begin 'evening discourses', to try to combat the danger of splintering into specialist groups. I was asked to give the first of these discourses, on 2 May, and I spoke about the Earth's atmosphere in the auditorium of the State Department, at the lectern where President Kennedy had, a year before, announced the decision to land a man on the Moon before the end of the decade. Kennedy himself did not participate in the COSPAR meeting, except to send a written message, but I did meet Vice-President Johnson. These geographical coincidences and chance meetings are deeply phoney, yet they did have the psychological effect of making me feel in the mainstream of advance in space in the real world, a feeling that faded in the 1970s and was utterly absent in the 1980s.

Another significant conference (to step forward in time for a moment) was in 1966, when the Royal Society asked me to organize a two-day international discussion meeting on the analysis of satellite orbits. I was able to invite six overseas speakers, and these included Fred Whipple, who spoke about the satellite tracking work at the SAO. The meeting, on 17–18 October 1966, had four main topics: satellite observations, determining orbits from the observations, and the two aspects of research from analysis of the orbits, on the upper atmosphere and the Earth's gravity field. I chaired some sessions, perched in the outsize red-cushioned President's 'throne' at the Royal Society in Burlington House, where Lord Blackett, the current President, had admitted me as a Fellow a few months before. This Royal Society meeting, the proceedings of which were published[38] in 1967, set the seal of respectability on orbit analysis, and successfully carried us through the next ten years, when the subject was declining in prestige just at the time when its scientific techniques were becoming more powerful.

To return to the early 1960s and to the RAE, my world of real work, James Lighthill continued as Director until the autumn of 1964. For the next five years the Director was Sir Robert Cockburn, who was personally interested in space research. During most of these years, however, the RAE was part of the new Ministry of Technology, and the emphasis was on diversification of the work into channels helpful for civil science. As most RAE work was either socially neutral or rather warlike, this fresh direction was exhilarating in that it might bring benefits to society.

The new Space Department formed in 1962 was not directly helped by the diversification trend, but still generally thrived during the 1960s under the leadership of George Burt, who succeeded Dr Lines as Head of the Department in 1963. In the mid 1960s the Department had about 120 scientific staff grouped in five divisions, covering satellite launchers (particularly the up-and-coming Black Arrow, discussed below), spacecraft (particularly the Ariel satellites, and later Prospero), engineering, instruments and electronics, and lastly the dynamics division, in which I was placed. The Head of this Division through most of the 1960s was Roy Bain, and our orbital dynamics group was one of five sections. Graham Cook and Harry Hiller were the two 'permanent' members of this section through the 1960s: Robin Merson, Bob Gooding and Russell Allan were in other sections.

The most visible work of Space Department was technological – the spacecraft design and engineering, and the satellite launcher work. The Department was involved in the design and construction of nearly all the British satellites of the 1960s and early 1970s: the scientific Ariel spacecraft (Ariel 1, launched 1962; Ariel 2, 1964; Ariel 3, 1967; and Ariel 4, 1971); the Skynet military communications satellites (Skynet 1A, 1969; the unsuccessful Skynet 1B, 1970, and Skynet 2A, 1974; and Skynet 2B, 1974); and the technological satellites Prospero (1971) and Miranda (1974). All these satellites were successful in design, though two of the Skynets went into the wrong orbits.

Even more demanding as an engineering project was the planned British satellite launcher of the 1960s, Black Arrow. In the late 1950s the RAE, with Westland Aircraft, had produced the liquid-fuel rocket Black Knight, intended to test the behaviour of ballistic missiles as they entered the atmosphere at high speed, and in particular to mimic the atmospheric entry of Blue Streak. Black Knight was successful in this role, but the cancellation

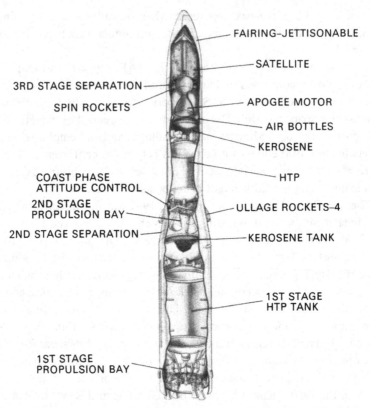

FAIRING–JETTISONABLE

SATELLITE

3RD STAGE SEPARATION

SPIN ROCKETS

APOGEE MOTOR

AIR BOTTLES

KEROSENE

COAST PHASE
ATTITUDE CONTROL

HTP

2ND STAGE
PROPULSION BAY

ULLAGE ROCKETS–4

2ND STAGE SEPARATION

KEROSENE TANK

1ST STAGE
HTP TANK

1ST STAGE
PROPULSION BAY

Fig. 4.14. Sectional drawing of the Black Arrow satellite launcher. The weight at launch was 18 000 kg, the maximum diameter 2 m and the height 13.2 m. A 100 kg satellite could be launched into a circular orbit at 500 km height.

of Blue Streak left it stranded like a whale. In the mid 1960s the decision was taken (by whom I am not sure) to build a satellite launcher at minimum cost by utilizing the proven Black Knight propulsion system of four Rolls-Royce Gamma rocket motors burning kerosene with hydrogen peroxide.

Black Arrow was the name given to this most economical of satellite launchers, and it is shown schematically in Fig. 4.14. The first stage of Black Arrow had eight Gamma motors arranged in four pairs and can be regarded as two Black Knights squeezed together, as it were. This first stage of Black Arrow had a diameter of 2 m, and carried about 13 000 kg of propellants, which burnt for about 130 s, giving a thrust equivalent to a weight of about 22 000 kg, about 20 % greater than Black Arrow's weight at launch of 18 000 kg. The second stage had two Gamma motors, and was thus in effect half a Black Knight, though the exit nozzles were enlarged because the second stage was operating *in vacuo*. The second stage had a diameter of about 1.5 m and carried about 3000 kg of propellants burning for about 140 s. After second-stage burnout, the second and third stages were stabilized in attitude until the orbital injection height was reached. Then the third stage, propelled by a Waxwing solid-fuel rocket, was spun up to a speed of 6 rev/s and ignited. After all-burnt, the satellite, still spinning, separated from the spent rocket, and both entered orbit. Black Arrow was launched from Woomera in Australia, as no site in Britain satisfied the safety criteria.

The first test flight of Black Arrow in June 1969 was suborbital by intention, but not completely successful. The second and third test launches were scheduled as orbital, and each was to carry a simple satellite called Orba – a thin-walled hollow sphere of diameter 30 inches, with a radio beacon. This project was suggested by Graham Cook, the idea being to determine the drag accurately by orbit analysis, and also, by making one sphere of aluminium and plating the other with gold, to evaluate the drag coefficient with two different surface materials. Sadly, neither of the Orbas (X1 and X2 as they were known) attained orbit, though the second very nearly did so in 1970, travelling from Woomera nearly to Alaska before entering the atmosphere. The next launch, of Prospero (X3), was successful – but belongs to the next chapter.

France had become the third country to launch a satellite, in 1965. In the late 1960s, with Black Arrow on course, it looked as though Great Britain might be the fourth. In fact we were to be pipped at the post by both Japan and China; but the prospect was enough to boost morale and keep up support for Space Department.

Though Space Department was working towards the birth of new satellites, the death of old ones was of more interest for orbit analysis, because we had never been able to analyse the air density at very low altitudes, just before the satellite's final plunge into the atmosphere. Few satellite observers had ever seen the fiery descent of a satellite at the end of its life, and in the 1960s special telegrams were sent out from Moonwatch to observers who might be well placed to observe such a spectacle. Unfortunately none of the decays was widely seen from Britain until 20 November 1968, when the rocket of Cosmos 253 came down over England. I was out watching, but by an unlucky chance – yes, my luck had begun to run out at last – the sky clouded over a few minutes before the satellite arrived. However, the firework display was seen from all over the country, and we were able to establish the satellite's track, which was published in

Nature[39] in a paper written with Doreen Walker and Pierre Neirinck, who both observed the event. Throwing out sparks of many colours, the satellite took about a minute to travel from over Lancaster (height 80 km) via Nottingham (70 km) to Southend (50 km) before finally burning out over the sea south of Dover. I have still never seen a satellite decay, despite the thousands of hours I have spent observing.

My memories of Space Department in the 1960s are coloured by the knowledge of its decline in the 1970s. What seemed normal then seems unattainable now. In particular, George Burt's achievement in keeping up the momentum now seems admirable, because the climate changed after his departure in 1968, though it might have done so if he had still been there. Among the memories of excellence in the 1960s, one of the strongest is the work of the Departmental librarian, O. W. Root, never known by his Christian name but always in the old style as 'Root'. He was unique among librarians because he actually read most of the reports he sent on to us and usually wrote witty or scathing comments on each. His remarks were often illuminating and had the good effect of making scientists look at the reports instead of ignoring them. Root was a chain-smoker who paid the penalty: he died in 1964, to the great regret of all who knew him.

The most efficient facet of my own work at the RAE – dictation – also began in the 1960s, and continued until my retirement. Although never allowed the luxury of a secretary, I was able to dictate letters daily, receiving excellent service from a succession of shorthand typists over more than twenty-five years. If I had to single out one, it would be Mrs Jeanne Godden, but they were all most efficient and helpful. Each gradually became familiar with my jargon of perigees, harmonics and strange symbols like Ω ('the horse-shoe'), and before long I would be taking unfair advantage by dictating technical papers. On an average day the dictation amounted to about four letters and a page (or half a page) of a scientific paper from a rough draft. In thirty years this produced about 25 000 letters (filed every six months and nearly all thrown away on retirement), and gave me well-typed drafts of the 200 or so papers of these years, drafts which were easy for me to amend and for others to read. These shorthand typists were in many ways my greatest benefactors at the RAE after 1960 because of their skilful daily help. They gave my correspondents the idea that I was highly efficient because answers arrived so promptly. The reality was the opposite: if incoming letters were allowed to accumulate, some would be mislaid, through *in*efficiency. I also owe a great debt to the mathematical typists in the Printing Department of the RAE: they converted the daunting handwritten originals of the papers on orbital theory into accurate typescript beautifully laid out. And the tracers wrought an equally miraculous transformation in our rough drawings.

Another great efficiency of the early 1960s arose from the kindness of the co-editor of *Nature*, Mr A. J. V. Gale, who printed all our papers very quickly without alteration and apparently without refereeing. In return I wrote book reviews for *Nature* whenever he asked, 42 in all during the 1960s, and many short news articles.

I steered clear of administrative responsibility at the RAE in the 1960s, and later. In the early 1960s my career figured in the Civil Service recruitment leaflet as 'Mr A', an example of rapid advancement. That had to be dropped because it was eight years before I was promoted in 1968 to Deputy Chief Scientific Officer in the individual merit scheme. As I remained in that grade until my retirement in 1988, I probably became the longest-serving *un*promoted DCSO in existence, 'Mr Z' rather than 'Mr A'. I rose like a rocket between 1954 and 1968, then fell from grace, stuck on a shelf.

I did undertake a few semi-administrative tasks in the 1960s. One, in 1964, was the writing of a report on how to write reports, at the request of the RAE Technical Publications Committee. This has now been through many editions, with necessary amendments in the rules but little change in the text.

James Lighthill was keen to increase the numbers of RAE Technical Reports and to have a greater proportion published externally. In 1964 each RAE Department was asked to set up a 'publications panel', to chase reports in preparation and to encourage external publication whenever appropriate. I became chairman of the Space Department Publications Panel, the only one that survived for long, and I remained chairman for nearly twenty years, during which time 500 Space Reports were issued, about half later being published in journals. I regard this as my chief contribution to the space literature: the Panel persuaded reluctant authors to write, forced lazy supervisors to release reports they were sitting on, and prodded authors to submit their reports to journals. I was a reluctant chairman, because I despised meetings as a waste of time. So I held the meetings at about 4 p.m. and, as 5 p.m. was going-home time, this concentrated the minds of the members wonderfully. They were usually the heads of Divisions, and they readily agreed to quite difficult 'actions' in chivvying recalcitrant authors and so on. Earlier in the day they might have preferred to spend time arguing against taking on such jobs.

Another semi-administrative success occurred when, as a member of the RAE Library Committee, I suggested a scheme for 'personal retention' loans of books, so that scientists could borrow on indefinite loan books certified to be specially important for them. Previously, a scientist had to scrabble to take notes before returning the book after a month's loan. (This was before the era of cheap photocopies.) I argued that books were comparable with laboratory equipment, which was not loaned but just supplied. This argument won the day.

I was lucky to avoid controversy in the 1960s, in view of my dabblings in the dangerous real world. The closest shave was over the US nuclear explosion in the upper atmosphere in 1962, Project Starfish, which damaged several satellites (including Ariel 1) by enhancing the intensity of the radiation zones round the Earth. I was a member of a committee on pollution in space chaired by J. A. Ratcliffe, by then Director of the Radio and Space Research Station at Datchet. The Government surprisingly accepted our report, which was printed as a White Paper (Cmnd 2029) in 1963. (Perhaps no one read it?) Previously I had been in trouble over the third (and last) article that I wrote for *The Round Table*, entitled 'Progress and Pollution in Space', which referred to the bad effects of the Starfish explosion. The Ministry refused to clear the paper for publication: I was not surprised, but Dermot Morrah was angry; and, as he seemed able to influence several members of the Cabinet (one of whom was his fag at Eton), the decision was miraculously reversed. It was a useful lesson in the power of 'the old-boy network'.

Perhaps this catalogue of mundane semi-administrative events ought to end with a dismal roll-call of the committees on which I served. But my eyes glaze over at the thought. To be fair, however, some of the committees were necessary, and all of them had the virtue of making me meet new people who were often helpful in the future. Otherwise, being so unsociable then, I should probably never have met anyone new, except at conferences. The Royal Society's committees, though sometimes tedious, I generally found quite pleasant and polite. My favourite was of course the Optical Tracking Subcommittee, which was so valuable in welding together the individual efforts of the visual observers. Optical Tracking's parent, the National Committee for Space Research, under Sir Harrie Massey, was also often quite lively (though not always).

The unpleasant committees were those run by the Science Research Council (SRC) (as it then was): scientists had to apply to them for grants – hold out the begging bowl, as I used to say. Having never then been reduced to such a state myself, I was appalled to discover the prevalence of this procedure. My naive faith was that a benevolent Director would direct me towards interesting projects, and that good work would be rewarded by a good salary. I preserved my naivety until about 1965, when reality broke in. My worst experience was as a member of the Space Policy and Grants Committee of the SRC: so much paper, such long meetings (always of men, of course), in that gloomy tower State House. Powerless to influence the system, I protested with a poem called 'Committee mini-men':

> Piles of paper, inches thick,
> miles of wordage, read it quick;
> or if you're sick of reading
> verbiage and special pleading,
> just stuff the papers in your case
> and start your journey, grim of face,
>   to the corridors of power
>   in the horrid Kafka Tower...

As the 1960s progressed, I became more concerned at missing life, pinned either in the ivory tower of theory and analysis or in Kafka Tower. Consequently I accepted an invitation to write a book called *The End of the Twentieth Century?* for a historical series being published by Macmillan. This occupied me much during the years 1967–69. The dangers of innate aggressiveness leading to nuclear or chemical/biological war, the prospect of famines, the environmental problems that became fashionable in the 1970s and 1980s, the likely advances in biology and technology – these and other such topics were covered. The book was quite successful, more than 30 000 copies being sold. But I had not submitted it for clearance, because it was not concerned with my work; this caused problems for the Ministry of Technology, then my employer, because I was commenting on some matters within the purview of the Minister, something a civil servant should not do. I had to plead guilty, but escaped retribution thanks to the intervention of Clifford Cornford, who was in a senior position in the Ministry.

Despite the problems it created, I did not regret writing the book, which led to new and wider intellectual fields. One such widening occurred during a family holiday at Portmeirion in August 1968, when I had a long talk with Bertrand Russell, who was then 96, at his house at Penrhyndeudraeth less than a mile away. I had admired Russell's work and style of writing since my schooldays, and at Cambridge I went to his lectures on philosophy. More recently, I had reviewed the first two volumes of his *Autobiography* for *Nature*. Now I wanted to ask whether he had changed his views on the likely future of the world since writing his book *Has Man a Future?* in 1961. The answer was 'no'; but I have given an account of the conversation in the journal *Russell*,[40] and will not repeat it here. Contrary to some reports, Russell was alert and witty, and not at all deaf. From his drawing-room he could see Shelley's white cottage at Tremadoc. I commented on this, and he spoke of Shelley with affection, saying he would have liked to write a book called 'Shelley the Tough', stressing his robust practical qualities. My book on Shelley had this bias, and Russell was amazingly keen to receive the copy I had brought for him. Later I found that the third volume of his *Autobiography* ended with a quotation from Shelley taken from my book.

Soon after finishing *The End of the Twentieth Century?* in 1969, I was asked to write three articles on space for a lavishly illustrated encyclopaedic publication, entitled *History of the 20th Century*, which was being issued in about 100 weekly parts. My contributions were in parts 88, 95 and 97. The first was entitled 'The beginnings of the space race' and covered the years 1957–60. The second was on 'Space in the sixties', with emphasis on the first landing on the Moon, on 20 July 1969. This was an event epoch-making enough to demand a few words more here, even though it had no direct effect on orbit analysis. One indirect effect was an invitation from Harold Wilson to a reception at 10 Downing Street in honour of the US astronauts. Among others invited were the players in the current television drama success, *The Forsyte Saga*: Neil Armstrong and Nyree Dawn Porter – it seems a nice epitome of the success stories of the late 1960s (though I cannot remember whether both of them were there).

My third article for the *History of the 20th Century* looked at the future in space, and included a whole-page illustration of what 'America's first colony on the Moon' might be like, the caption being 'Moon City, 1990'. Technologically the date was possible; if it now seems overoptimistic, perhaps that is an appropriate end to my voyage through an optimistic and forward-looking decade.

5

Into the realm of resonance, 1970–1979

Fertile valleys, resonant with bliss.

P. B. Shelley, *Queen Mab*

In 1970 a new world beckoned, the realm of resonance, with prospects of fresh and fertile fields of research. A satellite experiences resonance when longitudinal variations in gravity cause changes in the orbit that build up continually, day after day and month after month. Orbital changes that are basically very small then magnify themselves until they are large enough to be accurately determined: thus resonance creates a powerful technique for measuring the gravity field.

In earlier chapters the Earth's gravity has been taken to be composed of a series of zonal harmonics dependent only on latitude, and independent of longitude. This is an over-simplification, because in reality gravity varies with longitude: the variations are small, but detectable. The zonal harmonics discussed in previous chapters can be regarded as longitude-averaged, and each of them needs to be supplemented by a teeming family of harmonics that are dependent on longitude as well as latitude, 'tesseral harmonics' as they are called, after the tesserae of varied shapes in a Roman mosaic floor.

The variation of a tesseral harmonic with *longitude* is specified by its *order*. A tesseral harmonic of order 15 gives rise to 15 undulations as you go round the equator (or any other line of latitude), as shown in Fig. 5.1. The symbol m is used to denote the order of a tesseral harmonic: it is helpful to think of m as specifying the variations between one meridian and another. (The zonal harmonics, being independent of longitude, are tesseral harmonics of order zero.)

The variation of a tesseral harmonic with *latitude* is governed by its *degree*, usually given the symbol ℓ: it is useful to think of ℓ as specifying the

127

Fig. 5.1. The variation of geoid height with longitude created by a 15th-order harmonic. The diagram shows a slice through the equator, or parallel to the equator at any latitude.

variations between one *l*atitude and another. For any fixed value of m, a harmonic of degree ℓ and order m has $\ell - m$ zeros on going from south to north pole. In other words, there are *approximately* $\ell - m$ humps as you go along a line of longitude from south pole to north pole and then down to the south pole on the opposite side of the Earth.

Some numerical examples may help to clarify these numbing generalities. First, suppose that ℓ and m are both 4. There are 4 undulations with longitude, but no variation with latitude because $\ell = m$. It is like cutting an apple into four quarters with the stalk at the north pole, if you imagine each of the four sectors as having a slight hump on its outer surface. Another example is a peeled orange, having perhaps 12 sectors and closely mimicking the harmonics of order 12 and degree 12. These 'pure' sectorial harmonics only arise for equal order and degree. When the order and degree are unequal, the shapes become more complicated, and Fig. 5.2 shows the shapes corresponding to the harmonics of degree 8 and order 6, 7 or 8.

There is no need to take any notice of the detailed shapes in Fig. 5.2. The important thing to note is that the Earth's gravity field can be fully specified by a set of such harmonics, complete up to as high a degree and order as is feasible. The harmonics are defined in such a way that the degree ℓ must be greater than or equal to the order m. This halves the number of harmonics needed, but there are two numerical coefficients, written as $\bar{C}_{\ell m}$ and $\bar{S}_{\ell m}$, for each pair of values of (ℓ, m).† So a gravity field that is taken

† Mathematically, the gravitational potential of the Earth is written

$$\frac{\mu}{r} \sum \sum (R/r)^{\ell} P_{\ell}{}^{m} (\sin \phi) (\bar{C}_{\ell m} \cos m\lambda + \bar{S}_{\ell m} \sin m\lambda) N_{\ell m},$$

where λ is longitude, ϕ latitude, $N_{\ell m}$ a normalizing factor and $P_{\ell}{}^{m}$ the associated Legendre function. The summation over ℓ is from 0 to ∞, while m runs from 0 to ℓ.

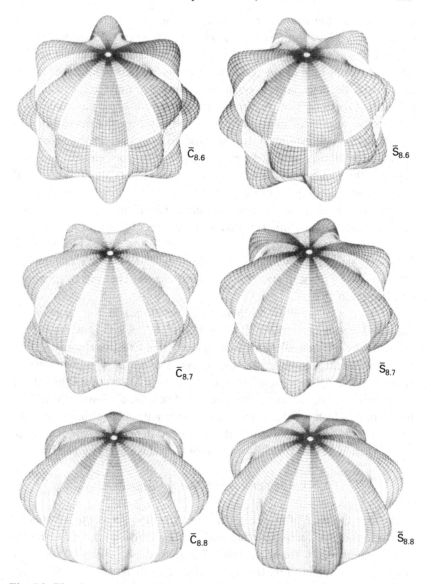

Fig. 5.2. The shapes produced by tesseral harmonics of degree 8 and order 6, 7 or 8, as computed by K. H. Ilk.

up to degree and order 36, as was typical in the 1980s, has 1296 coefficients waiting to be evaluated, no mean task.

An orbit is resonant when the satellite's track over the Earth repeats each day, and such orbits are uniquely sensitive to the harmonics of the same order as the resonance. Thus a satellite with a ground track that repeats every 15 revolutions (the tracks being 24° apart in longitude) reacts to the 15th-order harmonics of the gravity field. The effect builds up day after day, for as long as the resonance persists, and eventually the change in the orbit becomes large enough to be measured quite accurately, thus providing an estimate of the magnitude of the 15th-order harmonics.

When the numerical values of the 1296 harmonic coefficients in a 36 × 36 gravity field are determined simultaneously in a 'comprehensive solution', strong correlations arise between many of the coefficients, and this degrades the accuracy of the computation. Analysing resonant satellites of a particular order, say 15, was expected to – and did – give an independent and more accurate method for evaluating the coefficients of order 15 and degree 15, 16, 17, ... Similarly, analysing a number of 14th-order resonant orbits might give values of harmonics of order 14 and degree 14, 15, 16, ... More remotely, there was the possibility that 2-day resonances, e.g. 29 revolutions in 2 days, might serve the same purpose. That was the prospect opening up in 1970.

The theory of satellite resonance was developed in four papers[1] by Russell Allan in the 1960s and early 1970s. He showed how the orbital elements are affected when an orbit is nearly (or exactly) resonant. If an orbit contracting slowly under the influence of air drag remains nearly resonant for long enough, an appreciable change in several of the orbital elements should occur. Such a change would be most obvious in the orbital inclination, which is otherwise basically constant. Allan's theory showed that the rate of change of each orbital element depends on the resonance angle, written as Φ. For 15th-order resonance this angle is given by

$$\Phi = 360\, nt + \omega + 15(\Omega - v) \quad \text{degrees},$$

where t is the time after perigee passage (in days), n is the number of revolutions of the satellite per day, and v is the sidereal angle measuring the Earth's rotation – it rotates at 360.987 degrees per day. Exact 15th-order resonance occurs when $\dot{\Phi}$, the rate of change of Φ, is zero, namely when

$$n = 15.041 - 0.042\dot{\Omega} - 0.003\dot{\omega} \quad \text{rev/day},$$

where $\dot{\Omega}$ and $\dot{\omega}$ are in degrees per day. Thus the value of n at resonance does vary slightly with inclination, but is not far from 15 rev/day – it is 15.26 rev/day for inclination 45°, and 15.05 rev/day for inclination 90°.

As often happens in science, the first example of this type of resonance

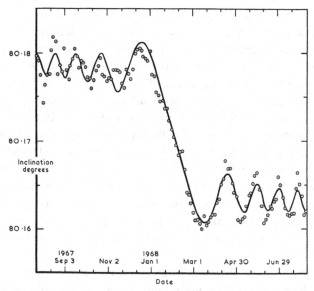

Fig. 5.3. Values of inclination for Ariel 3 at 15th-order resonance, with fitted theoretical curve.[2]

arose by chance, during a 'routine' orbit determination by Bob Gooding from Minitrack observations of the UK satellite Ariel 3, made for the benefit of the scientists whose experiments were aboard. On plotting out the values of orbital inclination, he found a remarkable decrease, between mid December 1967 and mid March 1968, of about 0.02°, equivalent to about 2 km in distance and much larger than the likely errors of about 100 m in the orbits. The decrease occurred at the time when the contraction of the orbit under the influence of air drag took the satellite through 15th-order resonance. Using Allan's equations, Bob Gooding wrote a computer program THROE (*t*esseral *h*armonic *r*esonance in *o*rbital *e*lements) which would adjust the values of 15th-order harmonics so as to give the best possible fit of the theoretical curve to the observed values of inclination. (Later THROE was extended to other orbital elements, and other orders of resonance.)

The fitting of Ariel 3 that emerged,[2] shown as Fig. 5.3, was completely convincing. Not only is the main decrease of 0.02° well modelled by the theory but so also are the subsidiary variations in inclination before and after the main effect. The rate of change of inclination due to 15th-order harmonics depends on the sine and cosine of the resonance angle Φ: near exact resonance, where Φ is nearly constant, the slope of the curve in Fig. 5.3 is nearly constant (and, as it happens, negative). As the satellite passes away from resonance, Φ changes and so do the signs of the sine and cosine terms, leading to the oscillations apparent in Fig. 5.3. Each full oscillation corresponds to a change of 360° in Φ. Well away from resonance, Φ goes through 360° in only a few days and the oscillation becomes very small.

The form of Fig. 5.3 is characteristic of most resonances that arise as an orbit contracts under the influence of air drag. The main change in inclination, centred on the time of exact resonance, may be either up or down, depending on the value that Φ happens to reach at resonance. (Sometimes the 'main' change is very small, if Φ takes an 'awkward' value by chance at resonance.) With a change of 2 km, as for Ariel 3, and an orbit good to about 100 m, it should be possible to evaluate the controlling 15th-order harmonics with an accuracy of about 5%. At that time the existing values for 15th-order harmonics were of very poor accuracy, about $\pm 50\%$, so the resonance analyses promised a great advance in knowledge of the gravity field.

The theory showed that the change in inclination at 15th-order resonance depended on the linear sum of a number of 15th-order coefficients, namely those of degree 15, 17, 19 and so on, of which the first two or three were likely to be the most important (at least for a near-polar orbit). One satellite like Ariel 3 would give a value for this linear sum, or 'lumped harmonic' as it was called: the lumped C harmonic determined by Bob Gooding was $(-19.9 \pm 1.2) \times 10^{-9}$. Ariel 3 was at 80° inclination: at other inclinations the lumped harmonic, though again a linear sum of coefficients of degree 15, 17, 19, ..., has different numerical factors. Thus each satellite would give a linear equation between the individual harmonic coefficients of order 15 and degree 15, 17, 19, ..., and these individual coefficients ($\bar{C}_{15,15}$, $\bar{C}_{17,15}$, ...) might be evaluated if we could analyse five or six orbits at different inclinations. It was essentially the same procedure as described in Chapter 4 for the zonal harmonics. This was to be the plan of campaign for the 1970s: we searched for satellites at different inclinations that had experienced resonance or would be doing so within a year or two, obtained orbits, analysed the resonance to determine values of lumped harmonics, and then evaluated the individual harmonic coefficients from the lumped harmonics at a number of different inclinations. The power of the method is shown by the accuracy of the lumped C-harmonic for Ariel 3: 1.2×10^{-9} is equivalent to an accuracy of 0.8 cm in geoid height, far better than could be achieved for 15th order by any other method.

Resonance also explained the strange decrease in the inclination of Transit 1B during 1962, and warned us that resonance might have affected some of the values of atmospheric rotation rate derived in the 1960s.

Fortunately, however, most of the satellites analysed for this purpose were of high drag. They passed through resonance so rapidly that there was no time for an appreciable change in inclination to build up. Transit 1B was a warning for the future: in deriving atmospheric rotation rates from changes in inclination, we would have to remove the effects of resonance, particularly if the orbits were of good accuracy and fairly low drag.

Although we had new territory to explore, it was with somewhat reduced numbers of staff. In the early 1970s Diana Scott left, and was not replaced, while Graham Cook was increasingly diverted to other work, and was lost to orbit analysis after about 1973. However, we were fortunate to be able to draw on the services of a new recruit, Alan Winterbottom, who was intended to assist Graham Cook but instead worked most effectively on orbit determination from 1970 until his transfer to another Division in 1974. Only Harry Hiller and Doreen Walker were officially working with me on orbit analysis throughout the 1970s, though we also received much essential help from Bob Gooding. Russell Allan concentrated mainly on orbital theory in the early 1970s and was later transferred to other work. He did not participate in the observational resonance analysis, for which his theory was the essential basis.

Though the siren sound of resonance was strong, the old territory of the 1960s was still asking to be explored. The exploration was conducted much more efficiently in the 1970s, because the new sources of observations led to much more accurate orbits. The procedure was to choose a good satellite, to gather observations over the whole lifetime (or perhaps just a year or two), to determine orbits at intervals of a week or two, and then to analyse these orbits for study of the atmosphere or zonal harmonics (and resonance too, if appropriate). For high-drag satellites the orbits determined from observations were much more accurate than any previously available, and much better results were obtained for atmospheric rotation rate and scale height.

The resonance analyses and the atmospheric studies both depended on the new sources of observations, the subject of the next section.

Observations new and old

The two new sources of observations that sharpened up the results of the research in the 1970s were the Hewitt cameras and the US Navy's Navspasur system. The other sources of observations – visual, kinetheodolite, Minitrack and Baker–Nunn – continued at about the same level through the 1970s, although the Baker–Nunn cameras were being used rather less. To the great regret of all visual observers, Moonwatch was closed down in 1975, a portent of further problems to come.

The two Hewitt cameras, purchased by the Ordnance Survey in 1967, came into regular use for satellite observing in 1968. The first camera remained on the existing site at Sheriff's Lench, near Evesham, and was operated by a team of eight, with Bill Sly as leader and Joe Hewitt available to deal with technical problems. The camera was devoted primarily to the campaign of the West European Satellite Triangulation, when cameras from various countries in Europe simultaneously observed the Pageos balloon satellite. As there was plenty of spare time between the successive transits of Pageos, the cameras also observed the satellites useful for orbit analysis, taking the priority list drawn up by the Optical Tracking Subcommittee.

The second Hewitt camera was installed at the Royal Observatory Edinburgh, or, to be more precise, at its Earlyburn outstation. A team of four observers, headed by Bennet McInnes and including Russell Eberst, operated the camera. But the observers had other duties too; so the camera was not as prolific as its twin at Evesham, though still most valuable, especially when both cameras observed the same satellite – the orbital accuracy then improved almost to the limit, which was about 5 m.

By 1975 the role of the Ordnance Survey had changed: selling maps was taking priority over geodetic science. It looked as though the Ordnance Survey would no longer need the cameras, and the operating team at Sheriff's Lench was gradually reduced. The camera at Edinburgh was also in decline by 1975 for different reasons. The Royal Observatory Edinburgh was part of the Science and Engineering Research Council (SERC) and, after the expansive euphoria of the late 1960s, the SERC was having to 'think the unthinkable' and make cuts in its expenditure on science. The Hewitt camera, an instrument only peripheral to the main astronomical work of the Observatory, was obviously vulnerable.

Another serious blow in 1975 was the death of Joe Hewitt at the age of 63, as he returned from a meeting of the Optical Tracking Subcommittee. He was my closest colleague among the satellite tracking community, and an ever-helpful friend in dealing with the organizational problems in national and international satellite observing. I greatly missed his friendly advice.

Orbit analysis in general was suffering a crisis in 1975. The main centre for the work was the RAE, which was part of the Ministry of Technology in the late 1960s, but was swallowed by the Ministry of Defence in 1971.

The problem arose because orbit analysis was non-military, and the responsibility for civil science funding lay with the SERC, which was supporting the prediction service; but no research with orbit analysis was being pursued in universities. This mismatch showed signs of being put right in 1974, when Dr Clive Brookes at the University of Aston in Birmingham put forward proposals for an Earth Satellite Research Unit at the University, to work on orbit analysis in close cooperation with the RAE. We were keen to encourage this expansion of the effort on orbit analysis, and, to cut a long committee-story short, Clive Brookes was soon able to set up the Earth Satellite Research Unit (ESRU) at the University of Aston, initially with two post-doctoral research associates, one of whom, Philip Moore, has continued to work most successfully on orbital research ever since.

By 1978 the Ordnance Survey was keen to find a new owner for the Hewitt cameras, and in September 1978 kindly handed them over to the University of Aston. The ESRU was then augmented by the addition of four observers, who also undertook orbit determination and analysis. The main problem with this new arrangement was the need for the observers to travel 40 miles to Evesham to operate the camera, a journey that was wasted if the prediction of a clear night was wrong, as often happened. The camera at Edinburgh was to be transferred to the southern hemisphere, to improve the geographical coverage; but the move did not occur until 1980 and belongs to the next chapter.

As a result of these various events, the observations from the two cameras were most plentiful in the early 1970s when the full Ordnance Survey team was in action; were falling away by 1975; reduced to zero for a time in 1978; and revived again in 1979, though only from one camera.

The second stream of new observations in the 1970s, from the US Navy's Navspasur system, had quite different origins. Navspasur is short for 'Naval Space Surveillance' and is a radar interferometer system. There is a very powerful transmitter at Lake Kickapoo in Texas, emitting an east–west fan-beam of radio waves. Any satellite crossing this east–west 'fence' reflects back some of the transmitted radiation, which is detected by six interferometer stations in a line across the southern United States, from San Diego in the west to Georgia in the east, at latitudes near 33° N. The Navspasur transmitter (Fig. 5.4) consists of 2256 dipoles stretching north–south for more than 3 km. The power radiated continuously is 810 kW on a frequency of 217 MHz (wavelength 1.4 m). The system became operational in 1961 and has been working 24 hours a day ever since, making about 10 000 observations of satellites per day.

Fig. 5.4. Aerial view of the Navspasur transmitter at Lake Kickapoo, Texas, looking south. The 2256 dipoles stretch for more than 3 km and the power radiated continuously is 810 kW. The lower half of the photograph covers only the first two of the eighteen sections: the rest stretch away into the distance and a public road (just visible) crosses the transmitter half-way along its length.

Dr Roger Easton, of the Naval Research Laboratory (NRL) at Washington, was the inventor of both Minitrack and Navspasur (and also of the Global Positioning System for navigation). When he visited the RAE in 1968, he knew that we were looking for reliable sources of observations; he also knew that the Navspasur observations were not being used for scientific analysis but only to help in producing NORAD orbits and in updating predictions. Roger Easton offered to send all the observations of any satellites we chose. This offer was eagerly accepted, and soon there was an agreed list of about forty satellites, mostly those already on the priority list for optical tracking. The list could be updated whenever we wished by correspondence with Roger Easton's colleague at NRL, Rudolph Zirm. This system worked admirably throughout the 1970s and 1980s, although both Roger and Rudolph retired about 1980.

These Navspasur observations provided the essential basis for nearly all our determinations of orbits by PROP for scientific analysis. The observations are unaffected by daylight, eclipse or bad weather and, for nearly all satellites, several observations per day have always been available, day in and day out, all through the year, for more than twenty years. The angular accuracy of the observations is similar to Minitrack or a good visual observation, about 0.03°. The Navspasur observations are all made from latitudes near 33° N, so the possibility of bias exists if there are no other observations. Usually, however, we had visual or Hewitt camera observations from other latitudes to counter this possible danger. In addition to the observations, we also received the US Navy orbits determined from the observations at the Navspasur control centre at Dahlgren in Virginia. These orbits, considerably more accurate than the NORAD orbits, were often used directly for analysis, particularly when we had no optical observations. So Roger Easton was a great benefactor of orbit analysis: without his inventiveness, foresight and kindness, much less would have been achieved in the 1970s and 1980s.

These two fresh sources of observations were very welcome, and so was a new twist in the tale of an old source, the kinetheodolite. Although only one kinetheodolite remained in operation through the 1970s, its geographical position gave it unique advantages. This was the instrument that had been at the Royal Greenwich Observatory in the 1960s: it was moved in 1969 to the South African Astronomical Observatory. This 'Cape kinetheodolite', as it was usually called, proved to be a prolific source of observations throughout the 1970s, thanks to the enthusiasm and hard work of Walter Grimwood. More than 15000 satellite transits were observed during the 1970s. Although the accuracy, about 0.02°, did not match that of the Hewitt camera, the kinetheodolite observations were in some ways more valuable, partly because they were more numerous, but chiefly because they 'tied down' the orbit in the southern hemisphere, where observations were otherwise rather sparse. (The most prolific and accurate visual observer in the southern hemisphere in the 1970s and 1980s was Bill Hirst, who moved to Cape Town on retiring from Moonwatch.) Adding Cape kinetheodolite observations to an orbit determination would sometimes improve the nominal accuracy of the eccentricity by as much as 5 times; and even when the accuracy did not improve, we always felt that the orbit was more reliable with these observations added.

In retrospect, visual observing in Britain seems to have flourished during the 1970s, apart from occasional threats to the prediction service. In 1972 Professor C. W. Allen retired after nearly fifteen years as chairman of the Optical Tracking Subcommittee and its predecessors. I was appointed as his successor, and I received such willing help from the many keen observers on the Subcommittee that the high profile of visual observing was kept up throughout the 1970s. About 20 000 visual observations were received each year, mainly from British observers, including large numbers from Russell Eberst, who remained the world's leading visual observer, and from David Hopkins at Bournemouth. The skill and knowledge of these and other observers on the Committee were vital in selecting satellites for the priority list, which was the key to future success.

The Subcommittee was able to organize four meetings of observers during the 1970s, in 1971, 1973, 1976 and 1978, all held on Saturdays at the Royal Society in London. All four meetings covered the techniques of visual observing and the research results obtained from orbit analysis; the last two meetings allowed the new University researchers to explain their plans for future orbit analysis, and to meet the observers. The general impression of 'progress', and the fresh faces appearing, certainly helped in boosting the enthusiasm of the established observers, as well as encouraging new ones.

The national satellite prediction service remained at Datchet through the 1970s, though the RSRS received a new name, the Appleton Laboratory, in 1972. During the 1970s Pierre Neirinck was synonymous with the prediction service, usually working or observing till late at night in the service of the satellites. His '24-hour service' was greatly appreciated by the observers, but not always by the administrators of the Appleton Laboratory. Because of the continual cost-cutting in the 1970s within the SERC, the prediction service was several times threatened with closure. I had to protest to the SERC on these occasions, and also endured some rather traumatic meetings of the Optical Tracking Subcommittee. Though closure was averted, the staff melted away, being reduced by 1979 to just one (or perhaps two – Pierre by day and Pierre by night). This was not quite as bad as it sounds, because the great advances in computers during the 1970s meant that the predictions could be produced with fewer staff. It was the 'public relations' aspect that suffered most. The attacks on the prediction service originated in the committees of scientists who advised the SERC – the 'committee mini-men' of the poem on p. 125. They wanted to see SERC money spent on 'good science'. The prediction service was not in itself good science: it was an 'enabling facility', a necessary preliminary to the observations that did lead to good science. Unfortunately it was in competition for the decreasing funds with projects that were expected to produce 'good science' directly. Hence the problems, which were alleviated for a few years by the new initiative at Aston University.

The closure of Moonwatch in 1975 was an unexpected blow to UK observers. Al Werner, Head of Moonwatch in its last few years, did his best to save it, but the trends of US science were against him. In the 1960s much of the American work in determining air density had been done at the Smithsonian Astrophysical Observatory by analysing orbits determined from observations made by Baker–Nunn cameras and (to a lesser extent) by visual observers. In the 1970s the upper-atmosphere physicists at the NASA Goddard Space Flight Center near Washington took charge, and they wished to pursue upper-atmosphere research by means of instruments aboard spacecraft, measuring the properties of the ambient atmosphere 'directly'. (This was partly because they were physicists and liked designing instruments rather than analysing orbits.) Whatever the reason, the Baker–Nunn cameras and visual observers were not supported. The use of the Baker–Nunn cameras for observing geodetic satellites was also becoming outmoded, because laser tracking was much more accurate – with errors in distance (by the mid 1970s) of only about 10 cm, as compared with positional errors of 10 m with the cameras. Some of the Baker–Nunn cameras were transferred to NORAD for use in tracking faint satellites and in improving orbital accuracy. The rest of the cameras were declared redundant, and Moonwatch was closed down. Most of the productive Moonwatch observers were transferred to the Appleton Laboratory's list, so the effect on observations was not too serious. It was the psychological damage that counted. Visual observing is an exacting and financially unrewarding task: observers need to feel they are wanted if they are to give their time and to brave the winter weather. The brutal message from the USA was that they were no longer wanted.

In the early 1970s we began to receive observations from the AFU-75 cameras operated by the Astronomical Council of the USSR Academy of Sciences. These observations were of excellent accuracy, about 0.002°, approaching that of the Hewitt cameras; but the satellites observed by the AFU-75 cameras were rarely those on the UK priority list, and their usefulness was limited for this reason. The USSR visual observations also continued to flow in, and were used whenever available for the appropriate satellites.

Observations from the Malvern radar were declining in numbers in the early 1970s, and the radar was eventually closed down about 1975. To compensate, the observations from Fylingdales, mentioned in the previous chapter, were becoming more easily available. By the end of the 1970s we were receiving data for several of the highest-priority satellites. These Fylingdales observations proved most valuable in orbit determination, because there were a good number every day, as with the Navspasur observations, and often on several transits.

With all these observations to hand, our orbit determinations became much more accurate and reliable than had been expected in 1970. By 1977, however, one tiresome defect was apparent – the inadequate numbers of observations during the last few days of a satellite's life, when the orbit changes rapidly and needs to be determined every day, or at least once every two or three days. Ideally we needed up to 100 observations per day, and there was only one possible source, the worldwide system of NORAD radars.

In October 1977, with this aim in view, Doreen Walker and I visited the headquarters of NORAD, inside Cheyenne Mountain, near Colorado Springs. This is a three-storey 'office-block' with a staff of nearly a thousand, and the building fits closely inside a huge cave carved out of Cheyenne Mountain: above the cave is nearly $\frac{1}{2}$ km of solid granite. To reach the NORAD headquarters, you have to drive along a tunnel about 500 m in length and through enormous steel doors that can be quickly closed. The buildings, made mainly of steel (with no glass), are mounted on huge spiral mechanical springs, about 3 m in diameter. (No, these are not the Colorado Springs). The Cheyenne Mountain complex is designed to keep going even after an attack by nuclear weapons: the air is filtered; communications are diversified; and food for a month is stored. The Cheyenne Mountain Complex is a potent symbol of our era, and will presumably become a fascinating 'ancient monument' if there are tourists in the years to come. Our meetings in 1977 were in the Command Room, which would serve as the national 'nerve centre' in case of missile attack – a reminder that working in 'the Mountain' was stressful, and could produce opposite extremes of euphoria, because it is so safe, or claustrophobia.

The idea of the visit was to obtain all the NORAD observations made in the last two weeks of the life of a satellite likely to be good for orbit analysis. NORAD agreed to this arrangement and, over the next few years, at our request, sent magnetic tapes with more than enough observations of about ten satellites. Excellent daily orbits were derived for all these satellites, allowing better determination of atmospheric rotation, density scale height and 16th-order harmonics. The results emerged in the 1980s and belong to the next chapter.

Computer programs for orbit determination and analysis

The observations were flowing in, and the computer programs were ready to process them. Bob Gooding's program PROP was in constant use through the 1970s for determining orbits from observations. The greatest virtue of PROP was its resilience in the face of orbits of very high drag: even satellites during their last day in orbit yielded to its gentle persuasions, thus extending the possibilities for upper-atmosphere research down to lower levels.

Having determined the orbit, we used the program PROD, developed by Graham Cook and Diana Scott, to remove the 'unwanted' perturbations due to the gravity field and lunisolar attraction. Then the orbits, cleared of these perturbations, were analysed, resonances being fitted with the aid of Gooding's new programs, not only THROE but also SIMRES, which could analyse inclination and eccentricity simultaneously. Atmospheric rotation rates were evaluated with a much simpler program called ROTATM.

The orbits determined by PROP were quite close to the Earth, usually only a few hundred kilometres up and much affected by air drag: at the other extreme, at a height of 36 000 km, were the drag-free geosynchronous satellites. The first such satellite in the UK Skynet series, for military communications, was launched in 1969, and Robin Merson had the responsibility for determining the orbit from radio observations. He and Alfred Odell developed the program SPOD, which proved more than equal to the task, and yielded orbits better than any previously obtained for geosynchronous satellites: the accuracy was about 10 m.

Between the geosynchronous and the low satellites are the high-eccentricity orbits with low perigee and very high apogee, including the numerous USSR Molniya communication satellites. PROD was essential for predicting the lifetimes of satellites in these orbits, for the Table of satellites: often the orbit had to be calculated for ten years or more ahead.

The pursuit of resonance

The analysis of resonance for Ariel 3 has already been shown in Fig. 5.3. The next step was to search for satellites at other inclinations that had passed through 15th-order resonance slowly enough to build up a substantial change in inclination. 'Slowly' means 'with low drag' and, since resonance occurs at a predetermined *average* height, the drag is lowest if the orbit is circular – otherwise perigee dips too low. So the search was primarily for nearly circular orbits, with eccentricity less than 0.01. Each new analysis would give a value for lumped (*C* and *S*) harmonics and consequently a linear equation between the individual (*C* and *S*) coefficients of order 15 and degree 15, 17, 19, ... With results from several different inclinations, there was hope of evaluating a number of these individual coefficients, and to do so more accurately than was feasible by other methods. This hope was to be fulfilled.

Fortunately there were two Russian satellites launched in 1970, namely Cosmos 373 at 63° inclination and Cosmos 387 at 74° inclination, that experienced resonance soon after launch in near-circular orbits of quite low drag. These were both analysed rather quickly by using US Navy orbits augmented by a few PROP orbits at times when Hewitt camera observations were available. The fitting for Cosmos 387 was excellent as far as it went, though a better result, shown in Fig. 5.5, was subsequently obtained from orbits extending over a longer time.

Armed with results from four satellites at different inclinations – namely Transit 1B at 50°, Cosmos 373 at 63°, Cosmos 387 at 74° and Ariel 3 at 80° – I offered a tentative solution for the individual coefficients of order 15 and degree 15, 17, 19 and 21, at the COSPAR meeting at Madrid in May 1972: it was published in *Nature*[3] two months later. The solution was tentative because it was 'high-risk', four coefficients being evaluated from four equations, with no redundancy. But time has shown that the results were good, and certainly much better than the values from the best comprehensive gravity fields that were available up to 1974, the Smithsonian Standard Earth III (SSE III, 1973)[4] and the Goddard Earth Model 6 (GEM 6, 1974),[5] which were both greatly in error, as is obvious from Table 5.1, where the best recent values are also given.

Opening up the new line of research on resonance was a useful boost for orbit analysis, because it promised an accuracy no other method could achieve – and also because it impressed scientific administrators who liked to snipe at the research for being 'in a rut' and confined to the techniques of the 1960s. In 1971 the Royal Astronomical Society awarded

Table 5.1. *Early values of 15th-order coefficients, from resonance and from comprehensive models*

Coefficient	Resonance[3] 1972	SSE III 1973	GEM 6 1974	Resonance 1989
$10^9 \bar{C}_{15,15}$	-18.3 ± 1.7	-56.4	-44.4	-20.4 ± 0.4
$10^9 \bar{S}_{15,15}$	-8.9 ± 1.6	34.9	35.6	-6.7 ± 0.4
$10^9 \bar{C}_{17,15}$	5.3 ± 2.1	49.1	not given	6.6 ± 0.5
$10^9 \bar{S}_{17,15}$	10.1 ± 2.1	32.0	not given	3.4 ± 0.6

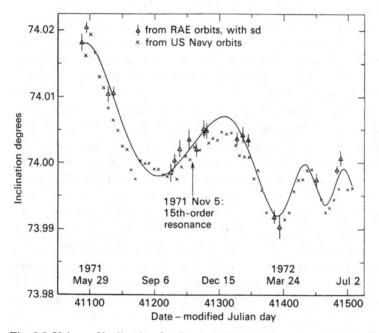

Fig. 5.5. Values of inclination for Cosmos 387 at 15th-order resonance, with fitted theoretical curve.[9]

me its Eddington Medal and also asked me to give the Harold Jeffreys Lecture. Entitled 'Heavenly harmony and earthly harmonics', the lecture reviewed the old problem of the spacing-out of the planets *à la* Kepler (who was born in 1571) and previewed the satellite resonance work then beginning.[6] It was a good chance to make scientists aware of the new technique.

Another conference on satellite geodesy was held at Lagonissi in May 1973. Some of the talks may have been obscure, but the sea was transparent and warm, and the weather perfect once again. During the conference Skylab was launched, and on its first evening in orbit I remember watching from Joe Hewitt's bungalow on the cliff top as the various pieces of the satellite passed over.

In 1973 the search was on for orbital data on past satellites that had experienced 15th-order resonance, and at Lagonissi Dr Fritz VonBun of Goddard Space Flight Center told me that much NASA orbital data was stored there, and invited a visit. In September Doreen Walker and I flew to Goddard (not in the first-class luxury of the 1960s, but in a stark and noisy military VC 10), and after also going to search the archives at the Smithsonian Astrophysical Observatory at Cambridge, Massachusetts, we returned with data on a large number of orbits that might be useful for resonance. At Goddard we met David Smith, by then Head of the Geodynamics Branch, and Carl Wagner, who was interested in resonance. There was also a visit to the Naval Research Laboratory to see Roger Easton and Rudolph Zirm.

The next six months saw analyses of the variations of inclination for eight of the orbits obtained, at inclinations between 31° and 90°. Adding three of the previous four gave a total of eleven pairs of equations for the C and S coefficients, and these were solved for eight pairs of coefficients, up to degree 31. The results were published[7] in 1974, and Table 5.2 gives the values up to degree 23, with recent values for comparison and also those from the GEM 8 model[8] produced two years later (the 1974 GEM model, GEM 6, only had values for $\ell = 15$, as Table 5.1 shows). Eight of the ten 1974 values in Table 5.2 are within 2 s.d. of the 1989 values, and the average difference is 1.4 s.d. The 1974 set of values is therefore quite reliable, as well as being surprisingly accurate, the error on the (15, 15) coefficients being equivalent to 0.6 cm in geoid height. In contrast, several of the values in the (later) GEM 8 model are seriously in error, though some are quite good.

So the analysis of inclination to evaluate 15th-order harmonics of odd degree can now be seen as an unqualified success: much easier than other methods, and much more accurate.

However, the fitting of inclination is not quite so simple as I have implied. The main effects are produced by harmonics of order 15 and odd degree, but the harmonics of even degree (16, 18, ...) are also sometimes influential, especially for orbits of eccentricity greater than 0.01. Also the 30th-order harmonics can be significant. In applying THROE, therefore,

Table 5.2. *Values of 15th-order coefficients of degree 15, 17,..., 23*

ℓ	$10^9 \bar{C}_{\ell,15}$ Resonance[7] 1974	GEM 8 1976	Resonance 1989	$10^9 \bar{S}_{\ell,15}$ Resonance 1974	GEM 8 1976	Resonance 1989
15	-21.5 ± 0.9	-24.0	-20.4 ± 0.4	-8.4 ± 0.9	-6.5	-6.7 ± 0.4
17	4.4 ± 1.6	2.4	6.6 ± 0.5	9.0 ± 1.5	18.3	3.4 ± 0.6
19	-15.6 ± 2.6	-8.8	-16.4 ± 0.6	-14.1 ± 2.7	-30.5	-14.2 ± 0.7
21	10.4 ± 3.0	27.2	18.3 ± 0.5	7.3 ± 3.5	13.8	12.0 ± 1.0
23	22.5 ± 2.8	17.5	20.6 ± 1.0	1.2 ± 4.4	-4.5	-1.4 ± 1.4

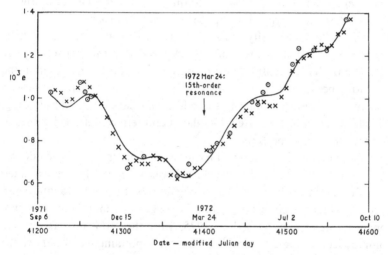

Fig. 5.6. Values of eccentricity for Cosmos 395 rocket at 15th-order resonance, with fitted theoretical curve.[10] ⊙ RAE orbits; × US Navy orbits.

there was always a choice of theoretical terms to be included in the fitting: we had to test out the effects of all likely relevant terms, and then to select the best group of harmonics for the fitting. Sometimes we determined fairly good values for lumped even-degree 15th-order harmonics or lumped 30th-order harmonics; but these were never so accurate as the primary lumped harmonics of order 15 and odd degree.

The lumped harmonics of order 15 and even degree, though having little effect on inclination, are the main agents in producing changes in orbital eccentricity and can be evaluated by analysing the changes in e at resonance for a number of satellites. My analysis in 1973 of the orbit of

Table 5.3. *Values of 15th-order coefficients of degree 16 and 18 from resonance and from comprehensive models*

Coefficient	Resonance 1975	GEM 6 1974	GEM 8 1976	Resonance 1989
$10^9 \bar{C}_{16,15}$	-13.7 ± 1.9	-47.5	8.7	-13.2 ± 1.2
$10^9 \bar{S}_{16,15}$	-18.5 ± 4.0	-37.8	-24.6	-26.5 ± 0.8
$10^9 \bar{C}_{18,15}$	-42.3 ± 2.7	not given	-61.4	-41.4 ± 1.3
$10^9 \bar{S}_{18,15}$	-34.7 ± 5.1	not given	-19.4	-17.2 ± 0.9

Cosmos 387 was the first to give a value for a lumped 15th-order harmonic of even degree,[9] with an accuracy of about 5%. Other analyses followed, and Fig. 5.6 shows a typical fitting, for Cosmos 395 rocket.[10]

In fitting the eccentricity, subsidiary pairs of harmonics sometimes needed to be added, and it was often tricky to decide the best set of pairs of coefficients. The eccentricity is much more difficult to analyse than the inclination, because the eccentricity suffers more serious perturbations from other sources, particularly the odd zonal harmonics and air drag. These effects must be accurately calculated and removed if good values of the harmonics are to be derived.

Soon we had values for lumped harmonics of order 15 and even degree from six satellites with resonant orbits. From these results we evaluated four pairs of individual harmonic coefficients of order 15 and even degree.[11] Table 5.3 gives the values for degree 16 and 18 in this first resonance solution, published in 1975, with the 50% increase in s.d. recommended. Table 5.3 also has the corresponding values from the Goddard models GEM 6 (1974) and GEM 8 (1976), and, for comparison, the 1989 resonance solution. Comparison with the 1989 values shows that the *C* values in the 1975 resonance solution are both excellent, but the *S* values in 1975 are not so good, being out by 2 s.d. or more: indeed GEM 8 is better for *S*. These rather poor values for *S* no doubt arose because the extraneous perturbations in *e* were not removed with sufficient accuracy.

When analysing the resonance of Cosmos 387 in 1973, I tried also to determine lumped harmonics from the changes in longitude of the node (odd degree) and the argument of perigee (even degree). Fittings were possible, but the results were considerably less accurate than those obtained from inclination and eccentricity respectively. Again the problem was the accurate removal of perturbations produced by the zonal harmonics. Thereafter we concentrated on analysing inclination and eccentricity.

By 1976, with reasonably good values for 15th-order harmonics available, our attention turned to 14th-order harmonics, to be derived from satellites that complete 14 revolutions while the Earth spins once relative to the orbital plane. These 14th-order resonant orbits have a much higher orbital period, about 103 minutes, corresponding to an average height of about 800 km above the Earth. The satellite passes through resonance most slowly if its orbit is near-circular: but there is so little drag on a near-circular orbit at 900 km height that it may take 50 or 100 years to come down to 800 km. Consequently, there were few circular orbits that had experienced 14th-order resonance: there were plenty of 14th-order resonant satellites in orbits of low perigee, but they shot through too quickly. Good 14th-order resonant satellites were therefore in short supply, and some of them had already been analysed by Carl Wagner at Goddard Space Flight Center, though he only used the inclination, not the eccentricity.

We managed to find four promising new orbits to analyse, and obtained good values of lumped 14th-order harmonics from all of them, using both inclination and eccentricity. To these we added lumped harmonics from the four orbits analysed by Wagner, and from four high-drag orbits, and derived values of the individual 14th-order coefficients of degree 14, 15, ..., 22. The results,[12] published in 1979, were not quite so accurate as for 15th order, but nearly so, the odd-degree values (from analysis of inclination) being better than the even-degree (from eccentricity). The values for degree 14–17 are compared in Table 5.4 with those of the Goddard Earth Model T3 (1992). Six of the eight resonance values are consistent with GEM T3. The striking exception is $\bar{C}_{14,14}$, for which the resonance value appears to be numerically too small, for unknown reasons. However, the remarkably low s.d. for the resonant value of $\bar{S}_{15,14}$ (it is equivalent to 0.2 cm in geoid height) may indicate that the GEM T3 values, though more recent, are not always superior to the resonance values.

There is nothing magic about 15 and 14: resonances of order 13, 12, 11, ... can be analysed too. Unfortunately, very few low-drag satellites experience these resonances, and the best specimen, Vanguard 3 at 11th order, was already being studied by Carl Wagner. Analysis of 16th-order resonance might also be possible, but as the average height of the satellite would be about 250 km, the very severe drag would pose serious problems. Consequently we did not try to grapple with 16th-order resonance until the 1980s; and orders 13, 12, 11, ... are still waiting.

More obvious than any of these is the first-order resonance exhibited by all the geostationary satellites, which make one revolution per day. These

Table 5.4. *Values of 14th-order coefficients from resonance (1979) and GEM T3*

ℓ	$10^9 \bar{C}_{\ell,14}$		$10^9 \bar{S}_{\ell,14}$	
	Resonance 1979	GEM T3 1992	Resonance 1979	GEM T3 1992
14	-38.5 ± 2.9	-51.8 ± 0.6	-7.8 ± 2.2	-5.0 ± 0.6
15	4.5 ± 1.1	5.4 ± 0.4	-23.8 ± 0.3	-24.3 ± 0.4
16	-22.3 ± 3.6	-19.7 ± 0.4	-36.0 ± 3.8	-38.7 ± 0.4
17	-15.0 ± 2.6	-14.1 ± 0.4	16.8 ± 1.2	11.6 ± 0.4

Table 5.5. *Some of Merson's values of low-order coefficients, and GEM T3*

Coefficient	Merson 1973	GEM T3 1992
$10^6 \bar{C}_{2,2}$	2.43	2.439
$10^6 \bar{S}_{2,2}$	-1.40	-1.400
$10^6 \bar{S}_{3,1}$	0.25	0.249
$10^6 \bar{C}_{3,3}$	0.70	0.720

satellites do not experience any drag, but this is of no consequence because the orbit is readjusted from time to time to preserve the resonance – otherwise the fixed aerials on Earth pointing at the satellite would have nothing to look at. Instead of continuous drag, the geostationary satellites are subjected to intermittent thrust. The drift in longitude of a geosynchronous satellite can be measured to give an observed value of a lumped harmonic which includes harmonics of several different orders, usually between 1 and 5. The most accurate geosynchronous orbits available in the 1970s were those of Skynet 1A determined by Robin Merson, and in 1972 he analysed the variations in longitude (between 40° E and 50° E) to derive appropriate lumped harmonics. He also produced an 'adjusted set' of individual harmonics of orders 1–5, which have stood the test of time quite well,[13] as Table 5.5 shows. The GEM T3 values are nominally accurate to 1 in the last figure quoted.

All the resonance analyses so far mentioned have relied on satellites that repeat their tracks over the Earth each day, after 15, 14 (or 1) revolutions. Was there any hope of analysing 2-day resonances, when the satellite makes perhaps 29 or 31 revolutions while the Earth spins twice relative to the orbital plane? The effects of such resonances on orbits would be much smaller than for a satellite of similar drag passing through 15th-order resonance, because 29th-order harmonics were expected to be only about a quarter as large as those of 15th order, and the change in the resonance angle would be twice as rapid. Thus satellites of low drag would be needed if the orbital changes were to be well defined. This pointed towards 29:2 resonance (average height about 650 km) rather than 31:2 resonance (average height about 350 km). But the number of suitable orbits has been disappointingly few. The first attempt at an analysis of a 2-day resonance[14] was in 1976 by Doreen Walker with Ariel 1. Unfortunately, the eccentricity was rather large (0.04), the perigee height consequently rather low (380 km), and the drag correspondingly high. As a result the values derived were not very accurate, but good enough to establish the feasibility of the techniques.

The 31:2 resonance was more promising in one way: the low average height ensured that the satellites encountering this '$15\frac{1}{2}$ resonance' would be more numerous than the 29:2 specimens. However, the lower height implied greater drag, and hence smaller effects. The first example arose by chance in 1976. In an earlier orbit determination of the Russian satellite Proton 4, Harry Hiller had found some anomalous changes in inclination, and we now realized that these changes occurred at the time of 31:2 resonance. The variation in inclination at resonance was analysed,[15] but the results were again not very accurate, because Proton 4 was a high-drag satellite (with perigee height 240 km) – the orbit determination had been intended for use in evaluating atmospheric winds. Again, however, the reality of the effect and the feasibility of the method were demonstrated.

These first studies of 29:2 and 31:2 resonance were particularly valuable for an unforeseen reason that had nothing to do with gravity. They showed the need to remove, or avoid, the perturbations due to both resonances, as well as 14th- and 15th-order resonances, when determining atmospheric rotation rates from changes in inclination for medium-drag satellites. In future, it seemed, the best plan would be to introduce 'natural breaks' as the orbit contracted through the 14:1, 29:2, 15:1 and 31:2 resonances, and to evaluate the atmospheric rotation rate in the intervals between these successive resonances.

In the late 1970s several orbits were selected that promised opportunities for analysis of 29:2 and 31:2 resonance, and the satellites were placed on the priority list for observing. In the 1970s Earth scientists were keen to have good values for harmonics of high order, such as 29 and 31: existing values were unreliable, and better knowledge of these harmonics would influence theories of the Earth's interior. But it was not until the 1980s that orbits were determined from the observations of the selected satellites.

The study of resonance is fascinating because it is like a mouse bringing forth a mountain. Harmonics of order 15 produce only very small 'hills and valleys' in the geoid, about 50 cm in height at most, yet can eventually generate large orbital changes. A typical change in orbital inclination produced by the 50 cm undulation is, in linear measure, about 1 m per revolution. So the change builds up at about 15 m/day, for as long as the resonance persists. With a resonance lasting for six months, as was typical for a low-drag satellite, the change would build up to about 3 km and could be quite accurately evaluated if the orbit was accurate to 50 or 100 m, as ours were. With a satellite of very low drag, an accuracy of 1 % might be possible – a change of 5 km measured accurate to 50 m. There was one actual resonant satellite, 1971–54A, which gave promise of such results, as it was taking several years to go through resonance.

By the 1970s there were not many other groups in the world working on orbit analysis, and only two scientists seriously pursued the study of satellites passing through resonance – Carl Wagner at Goddard Space Flight Center and (towards the end of the decade) Jaroslav Klokŏcník at Ondrejov in Czechoslovakia. We enjoyed the friendliest cooperation with both of them. There was plenty for everyone to do without any risk of rivalry: our forte was the analysis of 15th-order resonance.

In 1974 I was invited to give the Bakerian Lecture of the Royal Society, its premier 'prize lecture' in the physical sciences, endowed by the microscopist Henry Baker in 1774. The title of the talk was 'A View of Earth and Air'. The idea was to take a general look at the Earth and its atmosphere as seen by a visitor from another planet endowed with our current level of scientific knowledge. The lecturer's undoubted bias ensured that the results from orbit analysis were quite prominent, and included the values of the 15th-order harmonics given in Tables 5.2 and 5.3. The lecture, later published in the *Philosophical Transactions*,[16] was a good 'showcase' for the resonance studies, and helped maintain the public visibility that became so important for success in winning grants for university researchers in the tough times that lay ahead.

Other opportunities to present the results at conferences occurred from time to time, including a final visit to Lagonissi in May 1978; and this led in turn to an invitation to lecture at Delft in October 1978 from Karel Wakker, who was then beginning the orbital studies that flowered so impressively during the 1980s.

The resonance analyses took about half of our research time in the 1970s, but all other strands of the research were also flourishing, thanks to the much improved orbital accuracy made possible by the Navspasur and Hewitt camera observations. I shall discuss first the odd zonal harmonics, which in the phraseology of resonant orbits can be called 'the harmonics of order zero and odd degree'.

Refining the pear shape of the Earth

The values of odd zonal harmonics that we derived in 1969 were as good as was feasible at the time, but there were important gaps in the coverage of inclination. The first gap was the absence of near-equatorial orbits, the lowest inclination being 28°. By 1973 this gap could be partially plugged by results from three satellites launched in 1970, namely Explorer 42, Dial and Peole, at inclinations 3°, 5° and 15° respectively. The second gap was at inclinations near the critical value, 63.4°, where the oscillation due to odd zonal harmonics becomes very large. This gap was also partially filled after a three-year observation campaign by the Hewitt cameras and visual observers on Cosmos 248 at inclination 62° and Cosmos 44 at 65°. The orbits were determined at the RAE from these observations and the Navspasur data. For Cosmos 248 the oscillation in perigee distance had an amplitude of 29.6 ± 0.1 km, a larger value than any previously measured with such accuracy.

The five satellites mentioned above were added for a new determination of odd zonal harmonics,[17] published in 1974. Three of the previous 22 satellites were omitted and three others added, namely Explorer 32 at inclination 64.7°, Geos 2 at 105.8° and the British-launched Prospero at 82°. From these 27 orbits it was possible to obtain much improved values of the odd zonal harmonics: the nominal errors were reduced by factors of between 5 and 10 when compared with the 1969 set (2), see Table 4.2.

The values are given in Table 5.6 up to J_{11}, together with the sets obtained in 1973 by Kozai and by Wagner, and also those of the 1992 model GEM T3, for which the errors quoted are double the formal s.d. Our value of J_3 agrees well with GEM T3, being much closer than Kozai's or Wagner's. For J_5 and J_7, Kozai's values are in better agreement than ours with GEM T3; for J_9 ours is closest to GEM T3. Although most scientists would regard GEM T3 as the 'standard' because it includes much more data, it does not have any satellites near the critical inclination. So the superiority of GEM T3 is assumed rather than proved.

The pear-shaped Earth implied by our 1974 values is shown in Fig. 5.7.

Table 5.6. *Values of odd zonal harmonic coefficients determined by Kozai (1973), Wagner (1973) and King-Hele & Cook (1974), with values from GEM T3 (1992)*

Coefficient	Kozai 1973	Wagner 1973	King-Hele & Cook 1974	GEM T3 1992
$10^9 J_3$	-2541 ± 3	-2541 ± 12	-2531 ± 7	-2532.5 ± 0.4
$10^9 J_5$	-228 ± 4	-230 ± 13	-246 ± 9	-227 ± 1
$10^9 J_7$	-352 ± 7	-364 ± 16	-326 ± 11	-354 ± 4
$10^9 J_9$	-154 ± 7	-81 ± 23	-94 ± 12	-117 ± 6
$10^9 J_{11}$	312 ± 6	137 ± 34	159 ± 16	237 ± 9

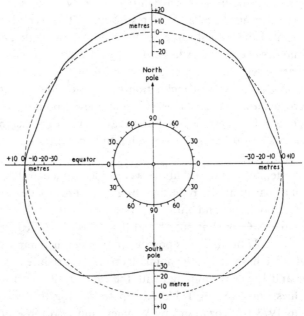

Fig. 5.7. Height of the geoid (solid line) relative to a spheroid of flattening 1/298.25 (broken line), as given by odd zonal harmonics[17] evaluated in 1974. Averaged over all longitudes. The geoid can be regarded as the mean sea-level surface.

The north polar hump is 18.9 m above the spheroid, and the south polar depression is 25.8 m. So sea level at the north pole is 44.7 m further from the equator than is sea level at the south pole. This profile remains within 1 m of that obtained in 1969 at most latitudes, but the pear-shape effect as measured at the poles (previously 40 m) changes by more than 4 m.

The pear-shaped Earth was still popular as a lecture topic. I gave a talk on the shape of the Earth to a large audience at Goddard Space Flight Center in September 1973; and when the International Union of Geodesy and Geophysics met at Grenoble in 1975, the President asked me to give the first all-Union lecture on the same subject. Here the audience was even larger, more than a thousand scientists ranging over the whole of geophysics. Afterwards the gist of the talk was published in *Science*.[18]

Is the upper atmosphere still going round too fast?

There had been some scepticism about our finding in the 1960s that the upper atmosphere at heights of 200–300 km was going round faster than the Earth. More accurate results for the rotation rate were therefore high on our list of projects. A number of satellites were selected for intensive observation because their orbits were suitable for determining atmospheric rotation. Most of the chosen orbits experienced 15th-order resonance as their orbits contracted, and at these times their inclinations suffered small changes – small because the drag was usually quite severe at this stage of the life. By 1974 we had fairly good values for the 15th-order harmonics, so the effects of the resonance could be removed with adequate accuracy. Nevertheless it was often more convenient – and convincing – to split the values of inclination at 15th-order resonance, so as to determine values of the atmospheric rotation rate at times before and after resonance. If the orbit had earlier experienced 14th-order resonance, a similar break would be made there. The effects of the 31:2 and 29:2 resonances were usually too small to be significant, but occasionally, as with Proton 4, a break was needed.

Our previous survey of upper-atmosphere rotation in 1969 had shown that the average rate of rotation Λ increased from about 1.1 rev/day at 200 km height to about 1.35 at 300 km. At that time Transit 1B had given an anomalous result; but after allowing for the effect of the 15th-order resonance, Transit 1B gave $\Lambda = 1.35 \pm 0.10$ at a height of 380 km and fitted into the picture much better.

The first modification of the 1969 picture came early in 1971, when analyses of three orbits at heights between 400 and 500 km indicated that Λ was lower in this region than at 300 km, probably being less than 1.0. This was reported in *Nature*,[19] with a graph showing Λ decreasing at heights above 350 km, to about 0.9 at 450 km. Two of the new satellites had circular orbits, and for these we used a new technique, plotting the inclination against orbital period rather than the normal method of

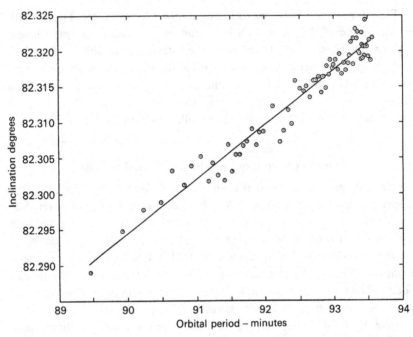

Fig. 5.8. Variation of inclination with orbital period[20] for 1963-27A, with fitted straight line.

plotting against time (as in Fig. 4.8 for example). For a circular orbit, the inclination should vary linearly with orbital period if Λ is constant, so the value of Λ is found by fitting a straight line, as in Fig. 5.8. The previous indication that Λ was greater in the evening than in the morning was confirmed by analyses of two further satellites: the winds were from west to east in the evening (say 18–24 hours local time) and, less strongly, from east to west in the morning (say 6–12 hours).

In 1976, after the accumulation of a few more results, particularly from Cosmos 54 rocket, we made another search for useful orbits, finishing up with 44 values of Λ that seemed reliable and accurate. The new orbits included Explorer 1 at the useful heights of 350–400 km; Ariels 1, 3 and 4 (400–500 km); the polar satellite ORS 2, which gave particularly good values; numerous Cosmos rockets in eccentric orbits; and several circular orbits.

The picture that emerged in 1976 was much more complex than before,[20] and is shown as Fig. 5.9. We divided the 44 points into three groups, morning (9 values), evening (8) and 'average' (27), and drew three separate curves through these groups. The division was obviously rather arbitrary

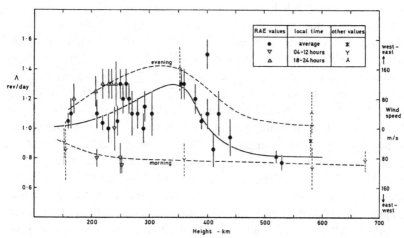

Fig. 5.9. The variation of atmospheric rotation rate Λ with height for evening, morning and average conditions,[20] based on 44 results.

and inexact, but was good enough to clarify the situation. The 'average' curve shows Λ increasing from about 1.0 at 150 km to about 1.3 at 350 km (which corresponds to a west-to-east wind of about 120 m/s), and then declining rapidly to about 1.0 at 400 km and 0.8 at 600 km. The upper broken curve shows that in the evening the wind is from west to east, increasing from about 100 m/s at 200 km to 150 m/s at 300 km, and then decreasing towards zero. The lower broken curve shows the morning wind as always east-to-west, and less strong. This new picture was a major advance, but there was still no good evidence of variations with season, solar activity or latitude. More values were needed to improve the picture and these were to emerge from several orbit determinations that were in progress in the late 1970s. However, these results were not brought together for a new synthesis until 1983, so they belong to the next chapter.

The idea of measuring meridional winds by satellite orbit analysis had been kept in mind for many years, but it was not until 1973 that a suitable satellite appeared, namely the final-stage rocket of Molniya 1-17. Such Molniya rockets quite often approach decay with a perigee height near 100 km and an apogee height that decreases by thousands of kilometres within a few weeks. Consequently they are very difficult to observe: the rapidly changing orbit spoils the predictions, which need to be updated daily. The visual observers in Britain made a splendid concerted effort to 'hold on' to this satellite in February 1973 and, aided by quite good weather, they succeeded, thanks mostly to the efforts of David Hopkins, David Brierley and Pierre Neirinck. From their observations and those of Navspasur, I was able to determine good orbits of this rapidly decaying satellite during the last 20 days of its life.[21] The eccentricity was large initially (0.3), and the other peculiarity was that the perigee position remained near the southern apex ($\omega \simeq 270°$). Atmospheric rotation has virtually no effect on the inclination of a high-eccentricity satellite if $\omega \simeq 270°$; so the changes in inclination are likely to be caused by meridional winds. I analysed the changes in both inclination and longitude of the node, and both agreed in indicating a south-to-north wind of 40 ± 30 m/s from 11 to 21 February 1973, a geomagnetically quiet time, and a south-to-north wind of 150 ± 30 m/s from 22 February to 3 March, a geomagnetically disturbed time. These results were for a height of 110–120 km, at latitude 63°–65° south and local time 6–12 hours. Such localized results at such a low altitude were something new for orbit analysis; but it was a feat that could not easily be repeated, as suitable satellites are rare.

Monitoring the density scale height

Most of the orbits specially determined from observations in the 1970s were chosen either because they experienced resonance or because they might give good values for the atmospheric rotation rate Λ. The resonant satellites were usually of low drag and did not tell us much about the atmosphere; but the Λ-satellites were of quite high drag. Consequently, their perigee heights decreased considerably during the lifetime, and values of the density scale height H could be found from the decrease in perigee height, in the manner described in Chapter 4.

The first orbit determination to yield a harvest of such values was that of Cosmos 54 rocket,[22] which gave thirteen values of H over its 5-year life at heights between 200 and 350 km. These agreed well with the new *COSPAR International Reference Atmosphere 1972* (henceforward called *CIRA*) and the only novel feature of the results was an indication that the scale height, and hence the atmospheric temperature, experienced a semi-annual variation, like the density. Accuracies of about 5 % in H were quite readily achievable; better accuracy could be secured only by taking longer time intervals, and thereby averaging out any variations due to solar activity, semi-annual effects, etc.

Several orbit determinations in the late 1970s gave good values of H. The best were in 1978, when the 100 orbits of Cosmos 462 determined by Doreen Walker and the 127 orbits of China 2 rocket determined by Harry

Hiller yielded a total of 48 values of H. A new feature of these results was the daily values in the last two weeks of the life, and their excellent accuracy: for example, $H = 24 \pm 1$ km at 195 km height from Cosmos 462. Rather boringly, however, these new values agreed well with the *CIRA* model. Most were within 10 %, and for Cosmos 462 the average difference between the observational values and *CIRA* was only 1 %.

These results established the techniques for determining scale height by orbit analysis, and it was good to know that the *CIRA* model could be relied on. There was no way of telling in advance that the work would turn out to be worthy rather than pioneering – monitoring the existing models rather than pushing into the unknown.

Air density and its variations

There was a danger that our evaluations of air density in the 1970s would also turn into a mere monitoring of *CIRA*. That did not happen because of the unpredictability of the semi-annual variation, but the studies of air density were rather pushed into the background in the early 1970s by the analyses of resonance.

There were, however, two extensive density determinations on low-perigee satellites begun before the resonance work erupted, and published[23] in 1971. The first of these gave 212 values of density at heights between 160 and 190 km, derived from the orbit of ATS 2 between July 1968 and September 1969. Despite its low perigee height, 180 km, this satellite remained in orbit for more than two years. The usual linkage of density with solar activity was evident, and the numerical values of density agreed quite well with those in current models. The semi-annual variation, however, was considerably greater than the models indicated for heights near 200 km, the maxima exceeding the minima by 32 % on average. The second of the two analyses was on Cosmos 316, a massive satellite with a perigee height of 154 km initially and a lifetime of 8 months, for which accurate values of perigee height were available because PROP was used to determine 36 orbits from observations. We derived 102 values of density, and again the semi-annual variation, this time in 1970, proved to be stronger than indicated by the models, with the maximum 30 % higher than the minimum, for heights near 160 km.

In the early 1970s there were also further studies by Graham Cook of satellites in high circular orbits, charting the semi-annual variation at heights near 1000 km over several years in the 1960s.

In 1973 Doreen Walker analysed the orbit of Cosmos 359 rocket and derived 145 values of density at heights near 200 km between August 1970 and September 1971, with accurate values of perigee height obtained from 42 orbits determined from observations.[24] The maximum densities in the semi-annual variation, in November and April, were on average 1.5 times the minima in January and July – again a much larger factor than indicated by *CIRA*. The semi-annual variation seemed to be changing from year to year, so its progress needed to be charted beyond 1971.

A major study on these lines was begun in 1976 by Doreen Walker, who determined 100 orbits of the satellite Cosmos 462 between December 1971 and April 1975, and then derived more than 600 values of air density over the $3\frac{1}{4}$ years, for heights between 210 and 250 km. These values had to be corrected to allow for variations in perigee height and the known effects of solar activity and the day-to-night variation (both of which are fairly small at this height). These adjusted values showed that the semi-annual variation was different in each of the years 1972 to 1975, but an averaged 'standard variation' could be defined, which was considerably stronger than in *CIRA*. If the average density through the year is taken as 1, the 1972–75 'standard variation' indicates a minimum of 0.94 about 24 January, a maximum of 1.28 about 5 April, a minimum of 0.77 about 21 July, and a maximum of 1.22 about 5 November.[25] This standard for the early 1970s has been used for various practical tasks, such as estimating lifetimes: it seemed more realistic than the *CIRA* model and also applied for heights (200–250 km) which are typical for the perigee height of a satellite due to decay in a few months.

This monitoring of the semi-annual variation in upper-atmosphere density was a good example of scientific work that is non-glamorous but valuable, just the thing *not* to win grants from research councils on the look-out for 'new and exciting science'. Yet how can you make comparisons between upper-atmosphere properties in July and October without knowing the average underlying change in density? It might be 20%; it might be 40%. Unless you know, you cannot assess the other effects you may be looking for. There was a chance to publicize the results at a symposium in Belfast in 1977 organized by Sir David Bates in honour of Marcel Nicolet's long career in aeronomy. But the resulting paper[26] was my last general survey of upper-atmosphere densities. In effect, the new vein of ore opened up in 1957 was being prematurely closed down, though much useful material remained, ready for later miners to extract.

During the 1970s, orbit analysis was being applied down to lower levels in the atmosphere. Two analyses continued until very near the end of the life, when the perigee height was below 120 km. The drag coefficient of 2.2

usually assumed for satellites with perigee heights of 150–250 km was based on the assumption of 'free-molecule flow': this applies if the mean free path of the molecules of the ambient air considerably exceeds the length of the satellite, as happens at heights above about 130 km. But the mean free path at a height of 120 km is only about 3 m, less than the length of some satellites. According to theory, the drag coefficient C_D decreases from 2.2 in free-molecule flow towards the value appropriate for continuum flow, usually near 1.0 but dependent on shape.

Did this decrease in drag coefficient really happen? Cosmos 316, which was about 4 m long, promised an answer, because a good orbit was obtained at a time 20 hours before the final descent, when the perigee height was 118 km. Analysis of the drag during the last day in orbit[27] showed that the drag coefficient decreased from its normal value, taken as 2.2 at a height of 124 km to about 1.5 at 119 km and about 1.2 at a height of 117 km. This was the first observational evidence of the decrease obtained from an orbiting satellite.

The other orbit yielding information about C_D was that of Molniya 1-17 rocket – which also gave the first good values of meridional winds. The perigee height decreased from 112 km to 108 km during the 20 days of the orbit determination, and direct values of C_D were found by taking the *CIRA* values of density. On the assumption that the mass of the satellite was 440 kg and the cross-sectional area $4.3 \pm 0.6 \, m^2$, the mean value of C_D obtained was 0.85 ± 0.20. This is lower than that expected from theory, 1.15 ± 0.1, but the error limits do just overlap.

The unusual orbit of this Molniya rocket allowed an unusual geophysical measurement to be made. The main change in the argument of perigee (about 0.5°) turned out to be caused by the oblateness of the atmosphere. Thus the ellipticity of the upper atmosphere could be independently measured. It proved to be equal to the Earth's ellipticity, 0.00335, with an accuracy of $\pm 20\%$. This numerical value was no surprise; but the capability of measuring it was unexpected, and created another new role in the repertoire of orbit analysis.

The underlying orbital theory

The basis for orbit analysis is the theory: without it, interpretation would be impossible. The most important work at the RAE in the early 1970s was the theory for the orbital resonances, developed by Russell Allan and, in its practical application, by Bob Gooding. This theory, already outlined at the beginning of this chapter, lay at the root of all the resonance analyses.

Further theory on the effect of the atmosphere was developed as necessary in the 1970s. The first necessary question was 'how much does the day-to-night variation in density affect the changes in inclination produced by atmospheric rotation?' We found that in special circumstances a 25 % effect was possible, but in most practical evaluations the effects were fortunately very much smaller and usually negligible.[28]

In the mid 1970s several promising satellites in orbits of high eccentricity were asking for their orbits to be analysed to evaluate the atmospheric rotation rate: unfortunately the only theory available was for small eccentricity. So a new theory was needed for eccentricities greater than 0.2. This was duly developed, and showed[29] that the change in inclination produced by atmospheric rotation depended on e through the function $(1-e)^{5/2}\ (1+e)^{-3/2}$. The change in inclination also proved to be proportional to $\sin i \cos^2 \omega$, as well as to the change in period, ΔT. Appropriate forms for meridional winds were also derived. These results contributed to the review of upper-atmosphere rotation summarized in Fig. 5.9.

Ever since 1957 we had been making predictions of the lifetimes of all satellites launched, for inclusion in the RAE Table of satellites. This work was gradually taken over by Doreen Walker in the 1970s. The predictions were based on orbital theory, with different techniques for different ranges of eccentricity, and with various corrections for the semi-annual variation, solar activity and (sometimes) atmospheric oblateness. The procedures needed systematizing, and in 1977 I tried to devise uniform graphical methods for calculating the lifetime L from the current decay rate $\dot{n}$, writing $L = Q/\dot{n}$. (Here $\dot{n}$ is the rate of change of n, the number of revolutions per day, so that the units of $\dot{n}$ are rev/day^2.) Then the idea was to produce graphs of Q that could be easily read for any value of e. Fig. 5.10 shows how Q varies with e: this diagram is useful for eccentricities greater than 0.2 (as long as lunisolar perturbations are not too large), but for smaller values of e the diagram needs to be greatly magnified for practical use. These larger-scale graphs, divided according to the level of solar activity, were given in the published paper,[30] together with methods for making corrections to the lifetime to allow for such effects as the semi-annual density variation. These methods were regularly used during the next ten years for making the lifetime estimates for all satellites launched.

Though the methods were abstruse, these lifetime predictions were 'real-world' events, because the satellites did come down through the lower atmosphere, always leaving a luminous trail (if seen at night) and sometimes also a heap of twisted smouldering metal on the ground. Usually these remnants of precision-engineering were scattered in the sea

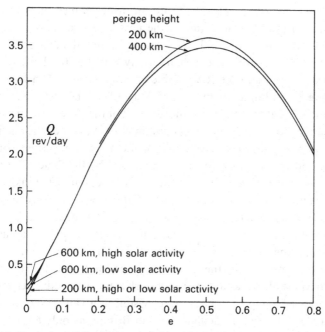

Fig. 5.10. Variation of Q with eccentricity e, for use in determining lifetime L from the decay rate $\dot{n}$ with the equation $L = Q/\dot{n}$.

or in areas remote from centres of population; but when pieces fell and were found, we were often asked to identify the satellite, generally a very easy task. Large satellites from which fragments might survive to reach the ground were entering the atmosphere at a rate of about two per week in the 1970s, but the news media took little notice of them until in January 1978 a Soviet satellite with a nuclear reactor as a power source, Cosmos 954, was placed in too low an orbit. Decay would be rapid and there was a danger that radioactive fragments would reach the ground. We predicted the decay date within the usual 10% accuracy limit through the last few weeks of the life, and, thanks to up-to-date observations from NORAD, our last prediction, about 24 hours before descent, gave the descent time accurate to 20 minutes. The satellite came down over northern Canada on 29 January 1978, and there was an expensive clean-up operation, though the hazards were less than had been expected because most of the radioactivity was dissipated in the atmosphere. This was one of the rare instances when orbital theory entered the public arena.

Another such event occurred in July 1979 when Skylab 1 ended its life in a blaze of publicity and fragments over Western Australia. The 75-ton

Skylab was launched in May 1973 into an orbit inclined at 50.0° to the equator at a height near 430 km. In the next 9 months it was occupied successively by three trios of US astronauts. After the last crew left in February 1974, the orbit decayed naturally and the total lifetime was predicted as six years in the RAE Table of Satellites issued in June 1974, implying decay within a few months of May 1979. The lifetime estimates made in the USA at that time took no account of the increase in air density as solar activity increased from its minimum in 1975 towards the maximum (expected about 1980). As a result, NASA stated that the lifetime would be ten years, and by then Skylab could be rescued by the Space Shuttle. In 1978 NASA realized that the satellite might decay in 1979, as we had predicted four years earlier. Some substantial pieces were likely to survive atmospheric entry, crashing to the ground at any latitude up to 50°. Most probably these would drop in the sea or in sparsely populated areas and, as no one on the ground had yet been killed or injured by man-made objects falling from orbit, there was no cause for alarm. But the media decided to create alarm, and there was a brisk trade in Skylab steel helmets. I found myself having to appear on television in a vain attempt to calm these fears, which were particularly silly in Britain, as only the tip of the Lizard peninsular was 'at risk'. During the weeks before decay we continually updated the predictions and, despite wide variations in solar activity, kept the error less than the statutory 10%. Ten days before descent our prediction was July 12.1 ± 1.0, and subsequent predictions were July 11.9, 11.7 and 11.6, with the date getting earlier because of increased solar activity. One day before decay, with the aid of recent observations from Fylingdales and other NORAD sensors, we were predicting descent at 14 ± 3 hours UT on 11 July; but on the morning of 11 July NASA manoeuvred the satellite slightly, extending the life by about one revolution, so as to avoid any danger of the satellite descending over the eastern USA. The actual descent was over the Indian Ocean and Western Australia at 16.30 hours UT. As expected, no one was hurt, but hundreds of pieces were found, the largest weighing about 500 kg. So orbital theory had come out of the ivory tower into the public eye again.

For some satellites in highly eccentric orbits the lifetime depends not on air drag but on the changes in perigee height produced by lunisolar attraction. By 1975 more than 40 Russian Molniya satellites had been launched into 12-hour orbits with eccentricities near 0.73, and these were much affected by lunisolar forces. The estimation of their lifetimes called for numerical integration with the program PROD. Wishing to avoid this brute-force method, I developed an approximate method for coping with the lunisolar forces, and showed[31] how the Molniya lifetimes depended on the initial orbital parameters, particularly the initial value of the longitude of the node Ω. For Molniya satellites at 65° inclination, as for most of those then launched, the lifetime was between 1 and 7 years, according to the initial value of Ω. If Ω was 250° initially, for example, the lifetime would be between 3 and 4 years; for $\Omega = 0$, the lifetime would be between 5 and 7 years. For Molniyas in orbits of 63° inclination, as in most subsequent launches, the theory was more difficult and the answers were less clear-cut. If the initial Ω was between 90° and 270°, the lifetime was likely to be between 9 and 13 years; for other values of Ω, decay is quite likely between 12 and 20 years after launch, but it is possible that the perigee may librate about the southern apex and, if so, no limit can be set to the lifetime.

The theory so assiduously pursued in the 1960s, specifying the contraction of orbits under the influence of air drag, suffered a hiatus during the 1970s, being stuck at Part VI: not until the late 1980s did any further Parts appear.

The world outside

In the wider world, the early 1970s marked the high point of British performance in space, the launching of the Prospero satellite by the British rocket Black Arrow on 28 October 1971. The launch, from Woomera in Australia, was near-perfect, Fig. 5.11. The 66 kg Prospero (1971–93A) entered an orbit inclined at 82° to the equator with perigee height 550 km, apogee height 1500 km and orbital period 106.5 minutes. The final-stage Waxwing rocket also entered orbit separately – and rather too enthusiastically, because it went on thrusting after separation and collided with Prospero, knocking off one of its four radio aerials. The other three aerials remained intact, and the asymmetry provided an unintended method of measuring the spin rate: also it was quite appropriate that Prospero should be attended in orbit by a (mis-spelt) Ariel. Though the free-flying aerial has now come down, both Prospero and the Waxwing rocket (1971–93B) are expected to remain in orbit for a hundred years. Will Britain have launched another satellite by then? Or is the UK in terminal decline technologically?

At the time of the Prospero launch we already knew that Britain was 'over the hill', because Black Arrow had been cancelled three months earlier. Cynics said it was to avert the danger that a successful launch might provoke demands for more. But there was also a more rational explanation. The space scientists serving on the SERC committees that controlled space research were not interested in Black Arrow. They were slightly

Fig. 5.11. The launch of Prospero by Black Arrow from Woomera on 28 October 1971.

embarrassed by the project, and would have preferred the money to be used for new spacecraft, innocently thinking that any spacecraft they were able to fund and build would, as in the past, qualify for a free launch by NASA. But once the British launcher was scrapped, the free US launches became a relic of the past. As Shakespeare's Prospero said, 'Our revels now are ended'. To go further and say 'Lord, what fools these mortals be' would not really be fair. The space scientists on these committees cannot be blamed for not foreseeing the 1973 oil crisis. The author of *The End of the Twentieth Century?* did not foresee it either. However, I did question the SERC's 'forward looks', which were always based on positive financial growth rates, such as 5% or 10%. 'What if it's zero or negative?' I would say, speaking the unspeakable to the annoyance of the bureaucrats. And the unspeakable happened, as so often in human life.

In the early 1970s all seemed well at the RAE. Euphoria still prevailed in Space Department, with the successful launch of Prospero in 1971 and the plans for further technological satellites to be launched by NASA, though in fact only one materilized, Miranda, sent into orbit by a Scout rocket in 1974. Sir Morien Morgan had returned to the RAE as Director in 1969 and stayed until his retirement in 1973: this was a fortunate appointment, as he had always supported the work on orbit analysis. But gradually the bad news began to overwhelm the good. The Ministry of Technology was abolished after a change of government in 1970, and the RAE was left rather in limbo, under the wing of a new Ministry of Aviation Supply. Black Arrow was cancelled in July 1971 – a portent of future cuts, as it turned out. Later in 1971 the RAE was taken over by the Ministry of Defence, and projects with a military flavour gradually began to take precedence over scientific work. The change was not dramatic, and no one knew that the trend would continue for twenty years. The new Director in 1973 was Rhys Probert, whom I had known in the 1950s while working on ramjet missiles. He did his best to support our researches, and indeed all the scientific research work of the RAE, until his unexpected death in 1980. But he was fighting against the tide in many ways. The RAE Director was now much more of a manager, conforming to the policy of the London Headquarters, which saw the RAE as becoming more concerned with supervising contracts with industrial firms and less with scientific research.

The Head of Space Department through most of the 1970s was Roy Bain. He did his best to combat the slow erosion of staff, but again he was fighting against the trend, because the Ministry of Defence was not then much into space projects. At the time, buoyed by the optimism of the 1960s, we were cheerful, believing that military applications of space technology would soon be recognized and supported. This process was slow, however, and there was a continuing decline in space efforts at the RAE during the 1970s and into the early 1980s. I was not involved in the administration of Space Department, though continuing as chairman of the Publications Panel (and still holding meetings at about 4 p.m.). Much of the Panel's work was done by the admirable Departmental Librarian,

Anne Maxwell, who knew more of what was going on than anyone else in the Department. I also continued on the RAE Library Committee and maintained friendly links with successive librarians. The RAE Library was, and still is, an excellent source of books and journals on aeronautics and space. Our research would have been far less effective without the instant access to the literature provided by the Library, and the subsidiary Departmental Library.

The RAE Table of Satellites remained in demand through the 1970s, and the monthly issues continued. By the end of 1979 the Table ran to 593 pages and was issued in three volumes with spiral binders. Producing the monthly issues and updating all previous pages was hard and skilful work, and throughout the 1970s the RAE was fortunate in having the services of Alan Pilkington, who had an annual contract to produce the basic entries for the Table, after being sent details of new launches by Harry Hiller, who then monitored the entries before they were typed, checked and sent out. The procedure usually worked well: although at times the Table seemed like a never-satiated monster, it was the best known RAE publication world-wide, so it seemed worth keeping up. An important feature of the Table was the lifetime predictions for all satellites.

In terms of staff numbers, orbit analysis at the RAE was in decline in the 1970s, but the quality of the work greatly improved, helped by the new observations, the PROP orbit determination program and better computers. The prediction service at Appleton Laboratory had been under threat because of staff cuts, but was surviving with fewer staff, as already mentioned. The orbital research also began to flourish outside the RAE with the creation of the Earth Satellite Research Unit in the Department of Mathematics at the University of Aston in 1974, and its expansion, and acquisition of the Hewitt cameras, in the late 1970s. Clive Brookes, Philip Moore and their colleagues were also contributing much to the researches on air density and the gravity field by 1978. The creation of ESRU was part of a larger project sponsored by the SERC, involving also the University of Leicester, where Professor Jack Meadows took the lead in starting work on orbits, with notable advances in orbital theory by Stefan Hughes and studies of satellite rotation by Vaughan Williams. The slow decline at the RAE was more than balanced by these new initiatives in the universities, and we cooperated closely with them. In 1979, after receiving an honorary DSc from the University of Aston, I was able to say that the Aston satellite project was flourishing, and that the sound of axes chopping had not yet been heard.

As a result of this strong university interest, there was more research with orbit analysis in the late 1970s than at any time before or since. More university researchers were recruited in these years, and their enthusiasm matured into useful results in the early 1980s.

A less successful venture in the 1970s was the plan for a laser satellite tracker in the UK. Satellites fitted with retroreflectors could be accurately observed at frequent intervals by the reflection of light pulses sent from a laser on the ground. At Goddard Space Flight Center, accuracies of a few centimetres were in sight in the mid 1970s, and attempts were in progress to monitor the movement of the San Andreas fault in California by this method. Many scientists in Britain were arguing in favour of a laser tracker, and none more so than George Wilkins of the Royal Greenwich Observatory at Herstmonceux, the obvious site for the laser. I was involved in various committee hearings, a Discussion Meeting at the Royal Society, and much correspondence. But the project never quite 'made it': there always seemed to be plans for expensive new spacecraft that appealed more to committee members (though the laser did materialize in the 1980s).

In the 1970s orbit analysis, including the new resonance techniques, came to maturity. More research was in progress and it was of higher quality, thanks to new observations and programs, better computers and better understanding. But the subject was no longer novel and trendy. By 1979 some scientific administrators tended to regard orbit analysis as 'old hat'. It was a warning that the 1980s would not be plain sailing.

For me home life, with the children growing up and books in gestation, seems more real and vivid than the trawled-up recollections of the last few paragraphs. The publication of *The End of the Twentieth Century?* in 1970 provoked many requests for lectures about the future, but these led to disillusion. There seemed no point in telling people bad things they did not want to know, for example about pollution, famine or chemical weapons. At one such talk, to a branch of the BMA, the air was so thick with cigarette smoke that breathing was difficult, and speaking even more so: every seat had an ashtray and few were unused. To tell the doctors that smoking was bad for them would get me nowhere. Though this example is extreme, I was also disappointed by the general attitude of futurists. Like most academic subjects, futurism provides intellectual exercise for its practitioners, and has only loose links with real life. (How many futurists predicted that revolutions throughout Eastern Europe would occur in 1989?) The question 'have we a future?' is taboo for futurists, who cannot be no-futurists. I aired some of these subversive grievances in an article published[32] in 1971.

It was a relief – escapism really – to return to the past, first to Shelley and later to Erasmus Darwin. Macmillan agreed to a second edition of the *Shelley*, which came out in hardback and paperback in 1972: it was good to have the chance to amend remarks I no longer approved of. I took a historical theme for the Halley lecture at Oxford in 1974 entitled 'Truth and heresy over Earth and Sky'.[33] The idea was to show how once-heretical ideas in geophysics and astronomy are now accepted, and vice versa. 'Always be sceptical' was the conclusion.

Erasmus Darwin was born again early in 1973 when John Scotney, a producer for BBC Radio 3, proposed that I write a 45-minute radio documentary drama about him. For me this was new and difficult territory; but eventually an acceptable script emerged, and was broadcast on Radio 3 under the title *A Mind of Universal Sympathy* in September 1973, with several subsequent repeats. Freddie Jones played Erasmus Darwin with memorable effect. I was more than grateful to John Scotney because Giles de la Mare, of the publishers Faber and Faber, heard the broadcast and asked me to write a full biography of Darwin. Away from the RAE the next three years were largely devoted to this book, which came out in 1977 under the title *Doctor of Revolution*. The reviews were excellent but the sales were not as swift as the publishers had hoped, and no paperback was issued. The book has now been out of print for several years, but finding a publisher for a second edition has so far proved impossible. The book did lead to two other projects, however. The first was another drama-tized documentary, entitled *The Lunaticks*, about the Lunar Society of Birmingham in the eighteenth century, broadcast in June 1978. The second and larger project was to find, collect and edit the letters of Erasmus Darwin, for publication by Cambridge University Press. It was a tough assignment, and I should never have started it without the help of Hugh Torrens of Keele University, who taught me how to identify obscure eighteenth-century people. Fortunately Darwin's letters are lively and amusing, so all the journeys scouring the country for possible manuscript letters seemed fascinating and worth while: the work continued into 1980.

6

On the shelf, 1980–1988

And when they were only half way up,
They were neither up nor down.

The Noble Duke of York

A rocket fired up the north face of the Eiger towards the summit might serve as a suitable simile for the worldly aspects of my career in science. From 1957 until about 1970 the upward thrust was strong, and the rocket seemed on course for the stratosphere. During the 1970s the propellant seemed to burn out and the momentum decreased. About 1980 the rocket came to rest on a rather precarious shelf, half way up the cliff: there was a danger of being pushed off into free fall; on the other hand, the position was a commanding one, from which good work might be done. As it turned out, the danger was averted and the decade was most productive.

In 1980 the researches based on orbit analysis seemed to be in good health. The Earth Satellite Research Unit at Aston University, under Dr Brookes, had moved to a spacious modern building at St Peter's College, Saltley, and the prediction service was transferred from the Appleton Laboratory to ESRU in July, because the Appleton Laboratory was being moved and merged with the Rutherford Laboratory. (Pierre Neirinck retired from Appleton but continued as a keen analyst of satellites.) In September 1980, when a meeting of visual observers was held at St Peter's College, ESRU was thriving, with four staff members working on predictions, four more as Hewitt camera observers, and a strong research team that included Philip Moore and two recently-appointed Research Fellows, Graham Swinerd and Bill Boulton, both working on orbit analysis and popularly known as the heavenly twins. The Evesham Hewitt camera continued in operation. The 'mothballed' camera at Edinburgh was being shipped to Australia, to be set up at the Siding Spring

Observatory in New South Wales. At Leicester University, Stefan Hughes had been awarded his PhD and was producing a stream of original papers on orbital theory and, later, on computer algebra for applying the theory.

The university euphoria was not matched at the RAE. The quality of the orbit analyses was still improving because of the better observations, more powerful computing facilities and deeper understanding of the theory. Doreen Walker was awarded her PhD in 1980 for a thesis on the analysis of Cosmos 462. But in 1980–81 the RAE suffered a cut in staff numbers. Among those who accepted the advantageous offers of early retirement was Harry Hiller: he retired in April 1981 after working with me for twenty-seven years and contributing much to the advances in orbit analysis in the 1970s. Fortunately Alan Winterbottom was allowed to return to work on orbit analysis and the Table of satellites. He continued to do so, most skilfully and effectively, throughout the 1980s.

As the trend at the RAE in the 1980s was away from research and towards monitoring contracts, the Table of satellites was in favour, because it involved a small contract, with Alan Pilkington until 1985 and then with Geoffrey Perry, for supplying the draft of each monthly issue. By 1980 the Table ran to 600 pages, and its printing (with monthly amendments), production and distribution was becoming onerous and expensive. Having contacts with Macmillan Press through my books, I asked whether they would consider publishing the complete Table as a book. Their editor of reference books, Rosemary Foster, was keen to do so. Accordingly we updated the Table to the end of 1980, involving about two months' work for Harry Hiller, Doreen Walker and, to a lesser extent, myself. The book was published in September 1981 as *The RAE Table of Earth Satellites 1957–1980* at a price of £30. It received excellent reviews: *New Scientist* called it 'a book that is stunning in its comprehensiveness', and the *Journal of the British Astronomical Association* said 'this book cannot be too highly recommended'. A second edition was soon requested by Macmillan. So we went through the lengthy updating process again at the end of 1982, with Alan Winterbottom working instead of Harry Hiller; *The RAE Table of Earth Satellites 1957–1982* appeared in September 1983. This sold out within about three years and another updating was undertaken at the end of 1986, with Doreen Walker responsible for the editing and Alan Winterbottom busy with the checking and the index (which ran to 42 pages). This heavy volume of 953 A4 pages came out[1] in 1987 at £65. The first edition sold 1511 copies, the second nearly 1300 and the third nearly 700, the total sales value being £150000. Not bad for a dull book! (Of course, the royalties went to the Ministry of Defence.)

While the Table thrived, the research only just survived. In the early 1980s there were grave doubts among the RAE management about supporting the research on orbit analysis – the stranded rocket was teetering on the edge of its narrow ledge. At a time of continuing cuts in, expenditure, orbit analysis was vulnerable because it did not contribute to the military projects that were now the main concern of the RAE. In 1982 the situation was particularly perilous, and was aggravated by quite separate problems caused by other cuts in funding, at the SERC and in the universities, which seemed to threaten the continuation of the Hewitt cameras. However, the RAE's attitude changed in August 1983, and an official work item was approved – the task of improving knowledge of the Earth's gravitational field by use of satellite observations. The reason for the change in attitude was never revealed, but I believe kind words from the military survey organizations in the UK and USA had some influence.

The change was linked with the inauguration of the satellite laser ranger at the Royal Greenwich Observatory, Herstmonceux, in 1983, because it might be possible to use observations by laser as well as those from the Hewitt cameras. On the initiative of Peter Whicher, then Deputy Director of the RAE, and with the approval of the Astronomy, Space and Radio Board of the SERC, a new body called 'The UK Facility for the Precise Tracking of Satellites' was set up in 1984 to oversee the satellite laser ranger at the RGO, the Hewitt cameras and the prediction service at the University of Aston. The Facility would receive funding in three equal amounts from the SERC, the Department of Trade and Industry, and the Ministry of Defence (in practice the RAE), and it was hoped that other agencies, such as the Natural Environment Research Council, would contribute (though in fact those contributions proved minimal). After the inaugural meeting, George Wilkins was the SERC representative on the management committee, and I was the MoD representative. The scientific programme for the laser was discussed at another committee with a formidable title, the Satellite Laser Ranger Users' Advisory Committee, SLRUAC for short. This had been set up in 1980, and I was its chairman for the first year or two before handing over to Professor Jack Meadows. The scientific programme for the Hewitt cameras was in effect decided by the choice of satellites for the priority list, drawn up by the Royal Society's Optical Tracking Subcommittee with the cooperation of the orbit analysts at the RAE and the universities.

The new funding arrangements seemed promising, but there were already problems at the University of Aston. The Vice-Chancellor in the late 1970s, Dr J. Pope, had been keen to take over the Hewitt cameras

Fig. 6.1. The Hewitt camera installed in the dome at the Royal Greenwich
Observatory, Herstmonceux, where it was operated from 1982 until 1990.

when the Ordnance Survey made their handsome offer. But the new Vice-
Chancellor appointed in 1980 had to face severe cuts imposed on the
University in 1981. He was not in favour of the work, and soon the Earth
Satellite Research Unit was moved back from St Peter's College to the
Main Building, where more rooms were available through redundancies
and early retirements – the staff of the Aston Mathematics Department
was eventually reduced from twenty-four to seven, and then merged with
other Departments.

Early in 1982 the Hewitt camera at the isolated site near Evesham
became electrically unsafe: this apparent further disaster proved to be a
blessing in disguise, because there was a large empty dome in the Equatorial
Group of telescopes at the RGO, Herstmonceux, and the camera was
moved there – a better home than it had ever known, with clearer skies
than at Evesham, the optical and technical expertise of the RGO at hand,
and the laser ranger only a kilometre away. Fig. 6.1 shows the camera at
Herstmonceux. Meanwhile the second Hewitt camera had been set up at an
even finer site – scenically and weatherwise – at Siding Spring Observatory

Fig. 6.2. View over Siding Spring Observatory, New South Wales. The Hewitt camera is visible on its concrete plinth to the left of the dome. The camera was operated here from 1982 until 1990.

in Australia, Fig. 6.2. The Herstmonceux and Siding Spring cameras both began full operation in 1982, and continued productively throughout the decade. For the first time there was the luxury of a camera in both northern and southern hemispheres, promising more accurate and more reliable orbits.

Despite the alarms and excursions of 1982, the orbit analysis work at the RAE continued with little decline in quantity and steady improvement in quality, leading to excellent results and several new advances in technique, as the next few sections will show.

Observations and orbits

Throughout the 1980s we continued to receive the Navspasur observations of selected satellites by courtesy of the US Naval Research Laboratory. As in the 1970s, these observations formed the essential basis for the orbit determinations, usually comprising more than 50% of the total – and sometimes even 100%, at dates when no other observations were available.

Fortunately it was also possible for Fylingdales to go on supplying daily radar observations of a small selection of satellites. Their observations contributed greatly towards the accuracy and reliability of these selected orbits.

The arrangements made in 1977 to receive all NORAD observations in the last two weeks of the life of specified satellites also continued in principle, and in practice too for a number of years, until difficulties arose in reading magnetic tapes because of changes in format. This was at the time when all the orbit analysis work was in jeopardy, and no effective action was taken, to my retrospective regret; so the arrangement lapsed by 1984.

During the late 1970s the most accurate optical observations of satellites of interest for orbital research had come from the Hewitt camera at Evesham and the kinetheodolite at the South African Astronomical Observatory. The kinetheodolite was by then more than twenty-five years old, and was wearing out, several parts having to be replaced. So it was planned to close down the kinetheodolite, and instead set up the second Hewitt camera in the southern hemisphere, as already mentioned. The kinetheodolite was finally closed down in 1981, after making more than 40 000 observations during its twelve years in South Africa. Those observations contributed vitally to the accuracy and immeasurably to the reliability of the orbit determinations, and of course we continued to use them throughout the 1980s in determining the orbits of satellites launched in the 1960s and 1970s. The improvement in reliability was never numerically measurable, because the nominal accuracy of the orbit, though usually improved by adding the kinetheodolite data, was sometimes worse. This happened when the orbit fitted to the northern-hemisphere observations was really 'overfitting' them, so that it became unrealistic in the southern hemisphere: the orbit was then 'brought to its senses' by a dose of kinetheodolite observations, the reliable anchor of its southern half.

The Hewitt camera at Siding Spring from 1982 onwards had an even stronger anchoring effect because of its better accuracy. After 1985 the southern camera was under the command of Rob McNaught, ably assisted by Bob McNaught – another (unrelated) pair of heavenly twins. During these years more than 3200 Hewitt camera 'plates' (actually films) were successfully exposed at Siding Spring in the tracking of close satellites. About five observations per plate were normally used when determining orbits, and these Australian observations contributed greatly to the accuracy and reliability of the orbits determined during the 1980s.

The Siding Spring camera was used for other purposes too. Was it possible to capture images of geostationary satellites, usually very faint, with magnitudes around 14? The answer was 'yes'. Some excellent plates were taken, each showing a number of geostationary satellites as dots, with the stars trailing as usual. The camera would have been well suited to monitoring these objects, had anyone wished to do so. Another unforeseen task for the camera arose from Rob McNaught's spare-time occupation as a discoverer of comets and novae, stars that suddenly increase in brightness. He made a crucial observation twelve hours before discovery of the famous 1987 supernova, the brightest for 300 years, and then by taking a Hewitt camera plate he identified the star that had 'flared up' so brightly.

The camera at Herstmonceux was also in full operation from 1982, with Max White in charge after 1985, assisted by Peter Strugnell. More than 3300 successful plates of close satellites were taken, again with about five observations per plate. Apart from its main programme of observations, the camera recorded at one extreme geostationary satellites and, at the other, the 'rogue' Russian satellite Cosmos 1402 very close to decay. The camera also helped the nearby laser ranger by providing independent checks on the angular accuracy obtained with the laser telescope and its capability in detecting geostationary satellites.

Throughout the 1980s there was no flagging in enthusiasm, productivity or accuracy among the volunteer visual observers who received predictions from, and sent observations to, ESRU. Each year the observers made about 20000 observations, which contributed greatly to the researches based on orbit analysis. Russell Eberst had made 100000 observations by 1984, observing mainly from his home in Edinburgh with 11×80 binoculars, and achieving an average accuracy of $0.03°$ in direction. David Hopkins in Bournemouth also reached a total of 100000 observations, in 1986, and showed great skill in tracking 'difficult' satellites. Among other skilled and prolific British observers in the 1970s and 1980s were David Brierley of Malvern, Peter Wakelin of Sunningdale and Mike Waterman of Camberley. The leading telescopic observer in Britain has been Gordon Taylor, who achieves accuracies between $0.01°$ and $0.02°$ and also specializes in observing geostationary satellites. All these observers, and others, achieve a very high standard of excellence – a higher standard in my opinion than in most professions.

Thus, for satellites launched after 1981, observations were usually available from Navspasur, the two Hewitt cameras, the visual observers and (for selected satellites) from Fylingdales. Some of the orbits determined in the 1980s, however, were of satellites launched in the 1960s or 1970s; for these, observations from the kinetheodolite, and the theodolite at Jokioinen, would also be included.

There were three further meetings of visual observers during the 1980s, organized by the Optical Tracking Subcommittee, of which I remained chairman. These helped to keep up the enthusiasm of the observers for their exacting voluntary task. The meeting in September 1980 at St Peter's College, Saltley, was one of the largest, and a group photograph is shown as Fig. 6.3. This was a two-day meeting (Saturday and Sunday), with a visit to the Hewitt camera at Sheriff's Lench on Sunday morning. The other two meetings, in March 1985 and September 1987, were both at Herstmonceux Castle, also over Saturday and Sunday. These were generally rated as the best of all the gatherings, for several reasons. The talks given by observers were of high quality; the rural environment was ideal; the RGO staff led by George Wilkins were most hospitable; the observers could visit the Hewitt camera and the satellite laser ranger; and in 1987 the Saturday night sky was superb, revealing the stars (and satellites) with a clarity that astonished observers who were accustomed to sites in towns.

By this time several observers had developed sophisticated methods for monitoring 'difficult' satellites, for which no good orbital elements were available. David Hopkins was keeping track of an extensive group of 'NOSS' ocean surveillance satellites, often 'recovering' them after a gap of as much as a year. Russell Eberst devised methods for identifying unknown satellites seen by chance. Pierre Neirinck followed the low-altitude US reconnaissance satellites which, although bright, often changed their orbits. Many other observers had their own specialisms; for example, Mike Waterman concentrated on new launches.

The procedures for orbit determination in the 1980s remained the same as in the 1970s, with PROP living up to its acronym as we leant on it heavily to produce accurate orbits for analysis. Inevitably, new computers came into use, and from 1985 onwards PROP was normally run on the RAE Honeywell machine under the 'Multics' operating system. As all the observations were stored on punched cards, we still sent in packs of cards to be read into the computer, an old-fashioned procedure that amused the RAE computer buffs but was still the easiest way to operate. Once the observations had been fed into the computer, we did have remote terminals from which the computations could be run, giving much greater speed and flexibility in orbit determination.

For resonance analysis the programs THROE and SIMRES continued in use, sometimes one and sometimes the other being preferred.

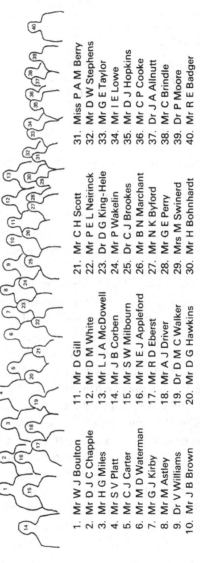

1. Mr W J Boulton
2. Mr D J C Chapple
3. Mr H G Miles
4. Mr S V Platt
5. Mr C J Carter
6. Mr M D Waterman
7. Mr G J Kirby
8. Mr M Astley
9. Dr V Williams
10. Mr J B Brown

11. Mr D Gill
12. Mr D M White
13. Mr L J A McDowell
14. Mr J B Corben
15. Mr S W Milbourn
16. Mr N E J Appleford
17. Mr R D Eberst
18. Mr A J Driver
19. Dr D M C Walker
20. Mr D G Hawkins

21. Mr C H Scott
22. Mr P E L Neirinck
23. Dr D G King-Hele
24. Mr P Wakelin
25. Dr C J Brookes
26. Mr B N Marchant
27. Mr N K Byford
28. Mr G E Perry
29. Mrs M Swinerd
30. Mr H Bohnhardt

31. Miss P A M Berry
32. Mr D W Stephens
33. Mr G E Taylor
34. Mr I E Lowe
35. Mr D J Hopkins
36. Mr C P Cooke
37. Dr J A Allnutt
38. Mr C Brindle
39. Dr P Moore
40. Mr R E Badger

Fig. 6.3. Participants in observers' meeting at St Peter's College, Saltley, Birmingham on 27/28 September 1980.
Photographer: Graham Swinerd.

Table 6.1. *Odd zonal harmonic coefficients from orbit analysis (1981),*
from GRIM3-L1 (1985) and from GEM T3 (1992)

Coefficient	Orbit analysis 1981	GRIM3-L1	GEM T3
$10^9 J_3$	-2530 ± 4	-2533 ± 2	-2532.5 ± 0.4
$10^9 J_5$	-245 ± 5	-241 ± 6	-227 ± 1
$10^9 J_7$	-336 ± 6	-337 ± 16	-354 ± 4
$10^9 J_9$	-90 ± 7	-143 ± 18	-117 ± 6
$10^9 J_{11}$	159 ± 9	233 ± 28	237 ± 9
$10^9 J_{13}$	-158 ± 15	-199 ± 42	-208 ± 14
$10^9 J_{15}$	-20 ± 15	-60 ± 44	-18 ± 18
$10^9 J_{17}$	-236 ± 14	-87 ± 48	-103 ± 18
$10^9 J_{19}$	-27 ± 19	30 ± 38	13 ± 17

A final polish for the terrestrial pear

Our previous evaluation of odd zonal harmonics in 1974, from twenty-seven orbits, did not have enough orbits at inclinations near the critical value, 63.4°. Two satellites of this kind, Cosmos 248 at inclination 62.2° and Cosmos 373 at 62.9°, had been high on the priority list for observing in the 1970s. Their orbits were determined by Clive Brookes at the University of Aston from 5000 observations – Navspasur, Hewitt camera, radar, kinetheodolite and visual – to evaluate the amplitude of the oscillation in perigee height caused by the odd zonal harmonics. For Cosmos 373, the amplitude was huge, 58.7 ± 0.2 km. In other words, the 40 m asymmetry between the northern and southern hemispheres affected the Earth's gravity field enough to create an orbital change of 58.7 km, measurable with an accuracy better than 0.4%.

Collaborating with Clive Brookes, and consulting Graham Cook (who was now doing very different work), I added new satellites to our previous data and gave what was for me a final polish to the values of the odd zonal harmonics. This solution,[2] published in 1981, included two new satellites, Cosmos 373 as already mentioned, and Explorer 46 at inclination 37.7°, filling a gap that existed previously between 33° and 41°. The new orbit of Cosmos 248 replaced the earlier one, which was over a shorter time. As one satellite was omitted (the least accurate of several at an over-represented inclination), the 1981 solution was based on 28 satellites. Three of the previous values for the amplitude of the oscillation were revised by fitting circles instead of sine curves.

The preferred new solution had nine coefficients (J_3–J_{19}) and the standard

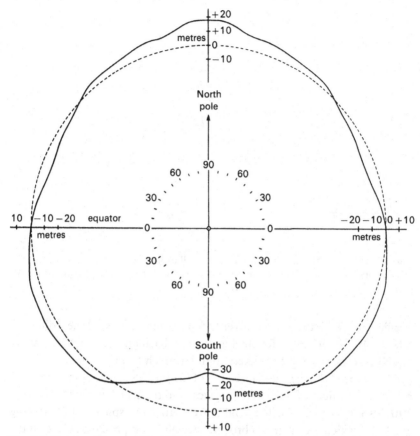

Fig. 6.4. Height of the geoid (solid line) relative to a spheroid of flattening 1/298.25 (broken line), as given by our 1981 values of odd harmonics,[2] with even harmonics of GEM 10B.

deviations were about 40% lower than in the previous (1974) solution. Table 6.1 compares the 1981 values with those in two subsequent comprehensive geoid models,[3,4] namely GRIM3-L1 (1985) and GEM T3 (1992), the s.d. for both these being doubled because they are intended as standards for comparison. The three values of J_3 are accurate and consistent. For J_5 and J_7, our values agree with GRIM3-L1 but are not consistent with GEM T3. Among the later coefficients there are significant differences, though all three sets indicate very small values for J_{15} and J_{19}, and large numerical values (greater than 100×10^{-9}) for J_{11} and J_{13}. There is no knowing which set is the best: the comprehensive models have the benefit of much more data, but they do not use orbits close to the critical inclination, which are particularly powerful in determining high-degree

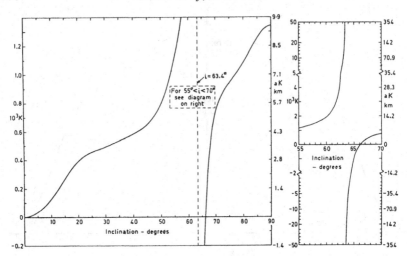

Fig. 6.5. Values of the amplitude K of the oscillation in eccentricity and of aK, the oscillation in perigee height, from the 1981 solution[2] for odd zonal harmonics (with $R/a = 0.9$).

coefficients. Perhaps the best verdict on Table 6.1 is, 'time will tell'. Obviously, better values for the high-degree harmonics are needed; but J_3 is well determined and is unlikely to differ much from -2.532×10^{-6}.

The shape of the meridional section of the Earth's sea-level surface given by our 1981 values of the odd harmonics (and the even harmonics of GEM 10B) is shown in Fig. 6.4. The heights are relative to a spheroid of flattening $1/298.25$, which is drawn as a broken line. With this profile, sea level at the north pole is 17.84 m above the spheroid, and sea level at the south pole is 27.23 m below the spheroid. So the pear-shape tendency – the north polar minus the south polar radius – is 45.1 m, as compared with 44.7 m in our 1974 solution. The profile of Fig. 6.4 should be accurate to 50 cm up to latitude 86°; over the remaining 0.2% of the Earth's surface very near the poles the errors may be larger, possibly up to 1.5 m. The basic reason for the increased error near the poles is that there is so little of the Earth's mass at these latitudes: thus a very peculiar shape very near the poles would have little effect on the orbit, and would not easily be detected.

The shape of a polar slice through the Earth is the most picturesque image of the effects of the odd zonal harmonics: but more important for orbital analysts is the size of the oscillation in perigee height produced by the odd harmonics. The amplitude of this oscillation needs to be accurately known because it has to be removed before an orbit can be analysed for other purposes (e.g. resonance). The amplitude of the oscillation in eccentricity (the quantity K defined on p. 111) and of the oscillation in perigee height (namely aK) is shown in Fig. 6.5 for inclinations from 0° to 90°: the scale on the left shows K, and the scale on the right gives aK. The diagram applies for $R/a = 0.9$, that is, $a = 7087$ km. (But aK is not sensitive to the value of a – indeed the value of aK for the Moon is near 3 km, quite similar to that for a close satellite of similar inclination.)

Fig. 6.5 shows that the amplitude of the oscillation in perigee height, aK, increases from zero for an equatorial orbit to about 4 km at inclination 30° and about 14 km for 60°, as shown in the separate diagram on the right. At inclinations just above 60°, aK increases further, to 59 km at 62.9° as Cosmos 373 revealed, and infinity at 63.4° (though it would take an infinite time to achieve this infinite value, and other perturbations would spoil the process, by altering the inclination). Normally, the perigee is closest to the Earth's centre when in northern latitudes. But, for inclinations between 63.4° and about 66.1°, the value of aK is negative: this means that perigee comes closest to the Earth's centre when in southern latitudes. For inclinations greater than 66° the value of aK becomes positive again, and increases continually, to a value of about 7.5 km at 75° inclination and 9.5 km for an exactly polar orbit.

These large orbital changes caused by odd zonal harmonics can have profound effects. For example, a 50 km decrease in perigee height may reduce a satellite's lifetime by a factor of 2 – or more if the original perigee is low.

Although our previous evaluations of odd harmonics, and pictures of pear-shaped Earths, had been presented at international conferences, this last and best effort was not. Money for conference-going was not easily acquired in the early 1980s because the orbital research was not fully approved within the MoD. The pear-shaped Earth had lost its magic: but the values of Table 6.1 still stand, not yet outdated, though probably destined soon to be so.

Resonance

In the 1970s we had pioneered the analysis of orbits passing through resonance. In the 1980s, having worked out the techniques, we were all set to make further advances. It was not easy because of the doubts about the future of the work, but progress did not slacken. In the 1970s we had derived good values for a number of individual harmonic coefficients of order 15, less good values for some of those of order 14, a few lumped harmonics of order 30, and some tentative lumped harmonics of order 29 and 31. In the 1980s all these existing results were greatly improved upon

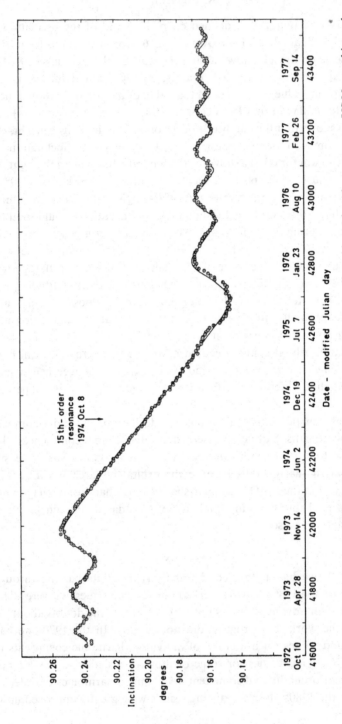

Fig. 6.6. The orbital inclination of 1971–54A during five years when it was strongly affected by 15th-order resonance: US Navy orbits and fitted theoretical curve.[5]

and new lines of research were opened, for example the analysis of 16th-order resonance. Progress with each order of resonance will now be discussed in turn, beginning with order 15, the flagship of the resonance fleet.

15th order

Three excellent analyses of 15th-order resonance in 1980 gave accurate new values for lumped harmonics of order 15 and 30 at three fresh inclinations. The first of the three used a satellite in a polar orbit, 1971-54A, that had been passing very slowly through resonance for several years. My analysis of 269 weekly US Navy orbits between November 1972 and January 1978 produced a 'copybook example' of the effect of resonance on inclination.[5] This is shown as Fig. 6.6, and has not yet been bettered by any other satellite. The fitting of the theoretical curve yielded values of lumped harmonics of order 15 accurate to about 2%, of order 30 accurate to 7%, and some tentative values for order 45. (Despite the excellence of this fitting, a slightly better one was achieved later by removing the small perturbation due to earth tides.) The fitting of eccentricity was not so perfect, but gave good values for even-degree lumped harmonics of order 15, the best being accurate to 3%.

In the other two new analyses of 15th-order resonances, the satellites were Intercosmos 11 at inclination 50.6° and Tiros 7 rocket at 58.2°. Doreen Walker analysed 112 weekly US Navy orbits of the former and 129 weekly orbits of the latter.[6] Again the variation of inclination with time was perfect in form for both orbits, as Fig. 6.7 shows. Although the time-intervals were shorter than for 1971-54A – being $2\frac{1}{2}$ years instead of 5 – excellent values were obtained for the lumped harmonics of order 15 and 30, the best being accurate to 2%. For these two satellites the fitting of eccentricity was also nearly perfect, as Fig. 6.8 shows for Tiros 7 rocket, the values of the lumped harmonics being accurate to 5%.

These encouraging results showed that analysis of slow resonances (of two years or more) would yield values of lumped harmonics far more accurate than achieved in current comprehensive models of the gravity field. At that time such models were being produced at the Goddard Space Flight Center (the GEM series), at Ohio State University (by R. H. Rapp and others) and by a European consortium of French and German scientists (the GRIM series). The GEM 10B, published[7] in 1981, was widely regarded as the best model of that time, and our resonance values provided a useful check on its likely errors, which we estimated as about 3×10^{-9} for the coefficients of order 15. On this basis, the value of the main

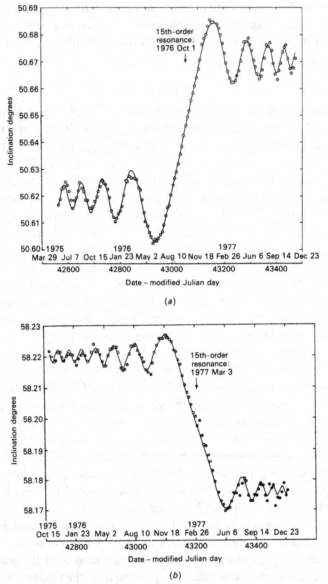

Fig. 6.7. The variation of inclination near 15th-order resonance, for Intercosmos 11 (a) and Tiros 7 rocket (b): US Navy orbits and fitted theoretical curves.[6]

lumped C harmonic ($\times 10^9$) for 90° inclination was -22 ± 4 from GEM 10B, whereas analysis of 1971-54A gave -16.4 ± 0.2: the nominal accuracy of the resonance value was 20 times better. Another way of looking at this

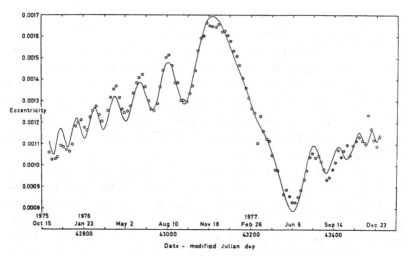

Fig. 6.8. Variation of eccentricity for Tiros 7 rocket near 15th-order resonance: US Navy orbits and fitted theoretical curve.[6]

accuracy is to think of the total undulation in the geoid due to 15th-order harmonics as being about 50 cm. Knowing the value of the dominant lumped harmonic with an accuracy of 2 % therefore implies an accuracy of about 1 cm in the contribution of the 15th-order harmonics to the geoid, whereas the overall geoid (which includes the effects of harmonics of all orders) was not known to better than about 2 m at this time. Resonance analysis gave such results quite easily. So it would have been a betrayal of science – with its aim of 'improving natural knowledge' – not to have pressed ahead with the researches, however lukewarm the official support in the early 1980s.

We soon used our new crop of lumped harmonics to derive a better set of values for the individual harmonic coefficients of order 15. As many other orbits as could be found were added in and, though none came up to the standard of the topmost trio, the gaps in the coverage of inclination were further reduced. This new solution,[8] published in 1982, made use of twenty-three different satellites at inclinations between 31° and 144°. Results were obtained for 15th-order coefficients of degree 15, 16, 17, ..., 35, though the accuracy was best for the lower degrees: the average s.d. for the coefficients up to degree 23 was 1.4×10^{-9}, equivalent to 1 cm in geoid height. A selection of the values obtained in this and later solutions is given in Table 6.2.

In this 1982 evaluation there was still a gap in the coverage of orbital inclination, because no accurate values of lumped harmonics were available for inclinations between 59° and 74°. The first step towards filling this gap was made in 1983 by an analysis of Cosmos 394 rocket at 65.8° inclination. Adding this satellite and revising some of the previous fittings, we redetermined the values of the individual 15th-order harmonics, again up to degree 35. This solution, published fully[9] in 1985, was more reliable – the uneasiness previously created by the yawning gap being alleviated – and also nominally somewhat more accurate. The changes in the numerical values were not startling, but often exceeded the previous standard deviations, as Table 6.2 shows.

In 1986 there were two further analyses of 15th-order resonance, mine of Tournesol 1 rocket at 46° inclination and Doreen Walker's of Cosmos 58, another satellite at 65° inclination that gave better results than Cosmos 394 rocket because of its lower drag. Armed with these new values, we derived a new solution for the individual harmonics of order 15 and degree 15–36, published[10] in 1987. Again there was a slight improvement in accuracy, but very few of the values changed by as much as 1 s.d. The average s.d. of the values for degree up to 23 in this 1987 solution was equivalent to 0.6 cm in geoid height.

Later in 1987 Alan Winterbottom determined 77 orbits of Cosmos 236, at 56° inclination, using about 70 Hewitt camera plates: this was the most accurate orbit that had as yet been analysed at 15th-order resonance. Also I determined the orbit of Pegasus 1, our lowest-inclination satellite (31°), to analyse the eccentricity and derive good values for even-degree lumped harmonics which had previously been lacking. In 1988 a further – and for me final – evaluation of the individual coefficients of order 15 was completed,[11] taking account of these two satellites and also using accurate values of a function that had previously been approximated. Again the solution was taken to degree 36, and there were great improvements – by a factor of about 3 – in the accuracy of the even-degree harmonics of degree 24, 26, 28 and 30, thanks chiefly to the new data from Pegasus 1. The other changes in this solution (published in 1989) were slight, as Table 6.2 shows. The values selected for inclusion in Table 6.2 (namely the coefficients of degree 15, 19, 20 and 28) are intended to be representative, rather than the best (which are of course those of lowest degree, 15, 16, 17...). GEM 10B is included as the best model of the early 1980s, with our estimates of its accuracy; GEM T3 is the best model available in 1992, with s.d. as given by the authors.

Table 6.2 shows that, for degree 15, the resonance values gradually improve in accuracy and do not alter much, though the change in C between 1982 and 1985 is substantial, being due to the accurate new value

Table 6.2. *Values of selected 15th-order coefficients in our four solutions in the 1980s, and values from two GEMs*

Coefficient	1982	1985	1987	1989	GEM 10B 1981	GEM T3 1992
$10^9\bar{C}_{15,15}$	-22.7 ± 0.6	-20.7 ± 0.5	-20.5 ± 0.4	-20.4 ± 0.4	-19.6 ± 3	-19.6 ± 1.0
$10^9\bar{S}_{15,15}$	-7.4 ± 0.6	-6.5 ± 0.4	-6.5 ± 0.4	-6.7 ± 0.4	-6.4 ± 3	-5.2 ± 1.0
$10^9\bar{C}_{19,15}$	-13.3 ± 0.8	-16.2 ± 0.7	-17.0 ± 0.6	-16.4 ± 0.6	-20.6 ± 3	-17.5 ± 0.7
$10^9\bar{S}_{19,15}$	-11.8 ± 0.9	-13.7 ± 0.6	-13.9 ± 0.7	-14.2 ± 0.7	-15.3 ± 3	-13.7 ± 0.7
$10^9\bar{C}_{20,15}$	-24.3 ± 2.3	-23.5 ± 2.0	-24.0 ± 1.3	-23.2 ± 1.1	-23.9 ± 3	-24.6 ± 0.9
$10^9\bar{S}_{20,15}$	-6.2 ± 1.6	-6.0 ± 1.5	-2.1 ± 1.0	-1.9 ± 0.9	4.8 ± 3	-1.4 ± 0.9
$10^9\bar{C}_{28,15}$	-15.4 ± 6.4	-15.2 ± 6.3	-19.7 ± 6.5	-10.6 ± 1.6	-6.8 ± 3	-10.9 ± 1.6
$10^9\bar{S}_{28,15}$	-8.4 ± 6.3	-8.4 ± 6.3	-6.5 ± 6.3	1.1 ± 1.4	-1.9 ± 3	-1.0 ± 1.6

Table 6.3. *Values of 15th-order harmonic coefficients of degree 15–23 as derived from resonance (1989) and in GEM T3 (1992)*

Degree l	$10^9\bar{C}_{l,15}$		$10^9\bar{S}_{l,15}$	
	Resonance 1989	GEM T3	Resonance 1989	GEM T3
15	-20.4 ± 0.4	-19.6 ± 1.0	-6.7 ± 0.4	-5.2 ± 1.0
16	-13.2 ± 1.2	-13.5 ± 0.9	-26.5 ± 0.8	-33.4 ± 0.9
17	6.6 ± 0.5	5.3 ± 0.6	3.4 ± 0.6	5.4 ± 0.7
18	-41.4 ± 1.3	-39.4 ± 0.9	-17.2 ± 0.9	-20.9 ± 0.9
19	-16.4 ± 0.6	-17.5 ± 0.7	-14.2 ± 0.7	-13.7 ± 0.7
20	-23.2 ± 1.1	-24.6 ± 0.9	-1.9 ± 0.9	-1.4 ± 0.9
21	18.3 ± 0.5	17.8 ± 0.8	12.0 ± 1.0	10.9 ± 0.8
22	23.2 ± 1.4	26.3 ± 1.0	6.7 ± 1.2	4.2 ± 1.0
23	20.6 ± 1.0	18.4 ± 1.0	-1.4 ± 1.4	-3.1 ± 1.0

from Cosmos 394 rocket in the gap between 59° and 74° inclination. For degree 19, a similar pattern prevails. For degree 20, the only significant change in the resonance value is for S between 1985 and 1987, caused by the inclusion of Cosmos 58: the accuracy improves by a factor of about 2 between 1982 and 1989. For degree 28, the big change is between 1987 and 1989, and simultaneously the s.d. is reduced by a factor of 4. Comparison of the 1989 values with GEM 10B suggests that the accuracies assigned to GEM 10B were quite realistic.

Comparison between the 1989 resonance values and GEM T3 calls for a more extensive listing, and Table 6.3 gives all the values up to degree 23. Generally the values in successive GEM models have come progressively closer to the resonance values as time goes on, and the GEM T3 values agree well with 'resonance 1989', apart from $\bar{S}_{16,15}$. If our standard deviations are realistic, those of GEM T3 seem rather overoptimistic, and this would not be surprising, as several values in preliminary versions of GEM T3 differed from the final version by more than 1 s.d. It will be some years before more accurate values can be established, so only time will tell whether our standard deviations are fully justified. On the assumption that they *are* correct, the 1989 results give excellent accuracy, offering a set of thirty coefficients of order 15 and degree 15, 16, 17,..., 29, which all have s.d. less than 2.0×10^{-9}. The average s.d. of these thirty coefficients is 1.3×10^{-9}, equivalent to an error of 0.8 cm in geoid height for harmonics of order 15.

30th order

When a satellite passes through 15th-order resonance, the 30th-order harmonics may be expected to have some influence, usually about 10% of the effect of the 15th-order harmonics. So, if the analysis is fairly accurate, with the 15th-order harmonics determined to better than ±10%, a closer fitting is likely if 30th-order harmonics are included; and determinate values of 30th-order lumped harmonics may well be obtained. The best such values are likely to emerge from analysis of the inclination, and they will be 30th-order harmonics of even degree, involving coefficients of degree 30, 32, 34, ...

By 1981, reasonably good values of lumped harmonics of order 30 and even degree had been obtained from seven satellites at inclinations between 50° and 98°. It seemed worth attempting a solution for individual coefficients, though the results would need to be treated with caution as there was a gap in the inclination between 58° and 74°. In this solution,[12] six coefficients were evaluated, up to degree 40; but only the 30th-degree values were well determined, with $10^9 \bar{C}_{30,30} = -1.2 \pm 1.1$ and $10^9 \bar{S}_{30,30} = 9.6 \pm 1.3$.

After this, the 30th-order coefficients were evaluated at the same time as the 15th-order. Thus new solutions were published in 1985, 1987 and 1989, in the papers already cited for 15th order. In 1985 only one new satellite was added, Cosmos 394 rocket, but the values for the 30th-degree C and S coefficients changed to -3.4 ± 1.1 and 8.3 ± 0.9 respectively. In the 1989 solution, from eleven orbits, the corresponding values were -3.2 ± 0.9 and 7.4 ± 1.0: the nominal accuracy of these 30th-degree values had not improved much, but that of the values for degree 32 and 34 was considerably better.

In 1988 Doreen Walker determined the orbit of Cosmos 184 rocket at 15th-order resonance from 3900 observations, including twenty-two Hewitt camera plates. On analysis,[13] this accurate orbit gave excellent values of lumped harmonics of order 15 and 30, and, as the latter seemed likely to have a strong influence, we made a final solution for 30th-order harmonics.[14] The standard deviations were 25% smaller than previously, and the average s.d. for degree 30, 32, ..., 40 was equivalent to an error in geoid height of 1.2 cm. The values are given in Table 6.4, with those from the GEM T2 model[15] and GEM T3 for comparison. The GEM T2 values differ greatly from those found from resonance and also have much larger s.d. The GEM T3 values have lower s.d. than GEM T2, and are generally closer to the resonance values, the average difference (2.6×10^{-9}) being

Table 6.4. *Values of 30th-order harmonic coefficients of even degree from resonance, with values from GEM T2 and GEM T3*

Degree l	$\bar{C}_{l,30}$			$\bar{S}_{l,30}$		
	Resonance	GEM T2	GEM T3	Resonance	GEM T2	GEM T3
30	-1.8 ± 0.8	1.4 ± 6	-0.1 ± 2.2	7.6 ± 0.7	0.5 ± 6	4.1 ± 2.2
32	-7.6 ± 1.6	4.5 ± 4	-4.3 ± 1.5	4.7 ± 1.6	-1.0 ± 4	-1.4 ± 1.5
34	-16.4 ± 1.9	-10.0 ± 4	-17.0 ± 1.3	-5.9 ± 1.8	-2.3 ± 4	-1.9 ± 1.3
36	-11.2 ± 2.6	-6.4 ± 2	-9.0 ± 1.1	4.8 ± 2.3	1.3 ± 2	4.3 ± 1.1
38	-0.9 ± 2.5	—	1.2 ± 1.4	3.0 ± 2.1	—	2.7 ± 1.4
40	0.0 ± 2.2	—	0.9 ± 1.3	-4.5 ± 2.1	—	1.1 ± 1.3

consistent with the s.d. Perhaps it is just wishful thinking to hope that future GEM models may be even closer. Or perhaps not!

So far, no solution has been published for 30th-order harmonic coefficients of odd degree derived from resonance: values of lumped harmonics are available at four inclinations (56°, 65°, 81° and 90°).

14th order

Our first evaluation of individual harmonic coefficients of order 14 was published in 1979. The results were quite accurate, but there were gaps in the coverage of inclination, which were partly filled by orbits analysed in 1984. The best of the new analyses[16] was that by Doreen Walker of Meteor 5 rocket at inclination 81.2°: this low-drag satellite remained close to resonance for four years (1977–81), and 208 weekly orbits were analysed to derive lumped harmonics of order 14 and 28, the best of which had an accuracy equivalent to 0.4 cm in geoid height. A new determination of individual 14th-order coefficients, based on fifteen satellites, was published[17] in 1986. We derived seven pairs of coefficients of odd degree, up to degree 27, and six pairs of even degree, up to degree 24. The results were a great improvement on the 1979 values, the five most accurate pairs having standard deviations corresponding to an accuracy of 0.9 cm in geoid height. The values up to degree 20 are given in Table 6.5, with GEM T3 for comparison.

As with the 15th-order coefficients, it is not improbable that the GEM T3 values should have rather larger standard deviations. Even so, the resonance values do not agree with GEM T3 as well as might be expected. In particular $\bar{C}_{14,14}$ is substantially larger numerically in GEM T3 (and

Table 6.5. *Values of 14th-order harmonic coefficients from resonance* (*1986*) *and GEM T3* (*1992*)

Degree l	$\bar{C}_{l,14}$		$\bar{S}_{l,14}$	
	Resonance	GEM T3	Resonance	GEM T3
14	-40.8 ± 1.2	-51.8 ± 0.6	-5.1 ± 0.9	-5.0 ± 0.6
15	4.6 ± 1.5	5.4 ± 0.4	-24.7 ± 0.8	-24.3 ± 0.4
16	-16.6 ± 1.6	-19.7 ± 0.4	-35.1 ± 1.9	-38.7 ± 0.4
17	-17.7 ± 2.3	-14.1 ± 0.4	17.9 ± 1.1	11.6 ± 0.4
18	-8.0 ± 3.8	-8.9 ± 0.4	-1.4 ± 2.7	-12.9 ± 0.4
19	-0.2 ± 1.8	-4.8 ± 0.4	-8.5 ± 0.7	-12.9 ± 0.4
20	12.2 ± 2.9	10.9 ± 0.5	-11.7 ± 2.5	-13.9 ± 0.5

most other comprehensive models), and the S coefficients of degree 17 and 18 are also in conflict. However, most of the other values agree well. If some of the resonance values are at fault (as seems more likely here than for 15th order), the probable reason is the lack of accurate 14th-order resonance analyses at inclinations between 50° and 66°.

28th order

The analysis of Meteor 5 rocket at 14th-order resonance gave reliable values of lumped 28th-order harmonics of even degree. By good luck these lumped harmonics depend primarily on the individual coefficients of degree 28, with a small contribution from those of degree 30, and negligible effects from higher-degree coefficients. With values from GEM T3 inserted for the small 30th-degree term, the lumped 28th-order harmonics found from the resonance of Meteor 5 rocket lead to the following values for the individual coefficients of degree and order 28:

$$10^9 \bar{C}_{28,28} = 8.3 \pm 1.2 \quad \text{and} \quad 10^9 \bar{S}_{28,28} = 3.1 \pm 1.2.$$

These compare favourably with the corresponding values from GEM T3, namely 7.3 ± 1.9 and 5.5 ± 1.9.

16th order

During the 1970s we were doubtful whether analysis of 16th-order resonance would ever be practicable, because the drag is so severe. At 16th-order resonance the orbital period is near 90 minutes and the satellite is usually within a few days of decay. The prospects improved, however, in

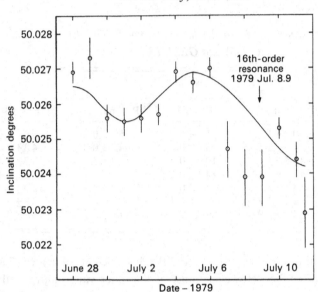

Fig. 6.9. Values of inclination for Skylab 1 in its last 15 days, with fitted theoretical curve for 16th-order resonance.[18]

1978 when we began to receive all NORAD observations during the last two weeks of the life for any satellite that we specially requested. Eventually five of these satellites proved useful in determining lumped harmonics of order 16 and odd degree (17, 19, ...).

The first analysis of 16th-order resonance resulted from Doreen Walker's accurate determination of the orbit of Skylab 1 during its last 15 days in orbit.[18] The inclination was the only orbital element that could be fitted, and the fitting was not so convincing as for 15th-order resonance. Fig. 6.9 shows the values, four of which had to be relaxed in accuracy, with the fitted curve. Quite good values of lumped harmonics were obtained, and subsequently we were able to make analyses of a similar standard on four other satellites, Cosmos 462, Intercosmos 10 rocket, Heos 2 rocket and NOAA B.

As these five satellites had inclinations of 50°, 65°, 74°, 90° and 92°, there was a reasonable spread, which encouraged us to solve for the individual coefficients. Solutions with three, four or five pairs of coefficients were obtained,[19] though only the first three pairs of values, of degree 17, 19 and 21, were determinate. Their accuracy was equivalent to between 1 and 3 cm in geoid height, and this was considerably better than the accuracy of the comprehensive models then available; these (and later models) rely mainly on terrestrial gravity measurements for 16th order. Table 6.6 gives the

Table 6.6. *Values of 16th-order harmonic coefficients from resonance (1987) and GEM T3 (1992)*

Degree l	$\bar{C}_{l,16}$		$\bar{S}_{l,16}$	
	Resonance	GEM T3	Resonance	GEM T3
17	-15.7 ± 3	-29.4 ± 1.1	4.2 ± 2	3.3 ± 1.1
19	-10.7 ± 4	-21.1 ± 1.0	-4.4 ± 4	-7.5 ± 1.0
21	17.7 ± 3	8.4 ± 1.0	-0.7 ± 4	-7.3 ± 1.0

resonance solution for three pairs of coefficients, together with the values now available from GEM T3. No one would advocate the analysis of orbits a few days from decay as a good method for evaluating gravitational harmonics: the set of values above is a *tour de force* rather than a serious challenge to the comprehensive models. Nevertheless, all the resonance values have the same sign as those from GEM T3, and two differ from GEM T3 by less than their s.d. Also it is again quite possible that the GEM T3 accuracies are overoptimistic.

11th, 12th and 13th order

Resonances of these orders are rarely encountered, except by high-drag satellites in eccentric orbits, which pass through the resonance too quickly for appreciable changes to build up. When a low-drag orbit does experience resonance, the variations of the orbital elements are richly detailed, with many 'side-resonances' occurring (for example when $\dot{\Phi} = 2\dot{\omega}$ or $3\dot{\omega}$), as shown in a classic study of Vanguard 3 at 11th-order resonance by Carl Wagner.[20] Because of the rarity of these resonances, it has not been possible to evaluate individual harmonic coefficients. All that can be done is to test the comprehensive models by comparing the lumped harmonics determined from resonance analysis with those given by the models. This should be possible in the next few years for orders 12 and 13 if the observations are maintained.

Harmonic coefficients of these and lower orders are of course evaluated in the comprehensive models of the gravity field with the help of much more accurate observations. The harmonics are determined primarily by measuring the strength of short-period perturbations experienced by the satellites of very low drag that are observed accurately by laser. 'Short-period' here usually means between 12 hours and a few days (or longer if

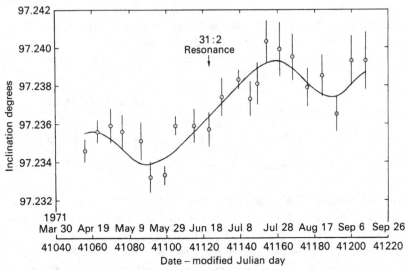

Fig. 6.10. Values of inclination for Samos 2 near 31:2 resonance, with fitted theoretical curve.[22]

the orbit is close to a resonance), and the effects are therefore small – whence the need for great accuracy. In our resonance studies we usually measured changes of 1 km or more, with the aid of orbits accurate to 50 or 100 m. The comprehensive models depend on measuring changes of a few metres with an accuracy of 1 m or better, relying on strength of numbers to improve the accuracy.

The two-day resonances, 31:2, 29:2 and 27:2

Now it is time to leave the fertile valleys of the one-day resonances and venture into the stony fields of the satellites that repeat their tracks every two days, after 31, 29 or 27 revolutions. There have not been enough of these analyses to allow sets of individual harmonics to be evaluated. The main value of the results has been to test the comprehensive models and assess their accuracy at these high orders (31, 29, or 27).

Progress was quickest with the 31:2 resonance, because more orbits experience it. Two good analyses were published in 1980, of Skylab 1 rocket[21] and of Samos 2[22]. The fitting of inclination for Samos 2 is shown in Fig. 6.10: the quality is not so good as for 15th order, but better than for 16th order. At the time, the lumped harmonics from these analyses were considerably more accurate than the values from the comprehensive models of the gravity field then available.

Table 6.7. *Values of harmonic coefficients of degree 32 and order 31, from resonance of Samos 2 and Cosmos 1335, and from GEMs T2 and T3*

Coefficient	Samos 2	Cosmos 1335	GEM T2	GEM T3
$10^9 \bar{C}_{32,31}$	-2.4 ± 1.4	-0.7 ± 2.0	-0.7 ± 6	-3.2 ± 1.9
$10^9 \bar{S}_{32,31}$	7.0 ± 2.3	9.5 ± 3.7	0.6 ± 6	0.6 ± 1.9

The only other good analysis of 31:2 resonance, of Cosmos 1335 at 74° inclination, was some years later.[23] The drag was higher than for Samos 2, so the results were not as accurate. However, it happens that the lumped harmonics from Samos 2 and Cosmos 1335 both depend primarily on the coefficients of degree 32. So, if values from GEM T3 are used to calculate the smaller contributions from the terms of degree 34, 36 and 38, both Samos 2 and Cosmos 1335 provide values of the coefficients of degree 32, given in Table 6.7, together with those from GEMs T2 and T3. The two independent pairs of values from resonance agree with each other to within half the sum of the standard deviations. This suggests that they are reliable and throws some doubt on the S value for GEM T3, which would give a curve very different from that fitted in Fig. 6.10. Again time will tell which pair of values is better.

Progress was slower with 29:2 resonance. Four analyses were available by 1981, but none was of satisfactory accuracy. The first good 29:2 analysis was made in 1983 when six lumped harmonics were evaluated from the changes in i and e at resonance of Cosmos 185 rocket at inclination 64°, using 83 US Navy and NORAD orbits.[24]

An even better analysis of 29:2 resonance, the best so far, was made by Doreen Walker a few years later,[25] of Cosmos 220 rocket at 74° inclination: 74 orbits between January 1983 and July 1985 were determined from 6300 observations, Hewitt camera observations being available for 38 of the orbits. The variations of inclination and eccentricity at resonance were fitted with theoretical curves to evaluate six lumped harmonics of order 29, the two of even degree having accuracies equivalent to 0.3 and 0.6 cm in geoid height, much better than in any previous attempt. These and the other four lumped harmonics were in quite good agreement with the values obtained from currently available comprehensive models.

By good fortune, the most accurate lumped harmonics derived from this satellite depend primarily on the individual coefficients of degree 30: those of degree 32 are negligible, and the contributions of those of degree 34, 36 and 38 are small enough to allow the GEM T3 values to be used. Thus Cosmos 220 rocket gives values for 30th-degree coefficients, as follows:

$$10^9 \bar{C}_{30,29} = 1.1 \pm 0.8 \quad \text{and} \quad 10^9 \bar{S}_{30,29} = 6.8 \pm 1.2.$$

The corresponding values from GEM T3 are 2.2 ± 1.7 and 3.3 ± 1.7. So those obtained from resonance are nominally the more accurate.

In the lumped harmonics derived from analysis of Cosmos 185 rocket, the terms of degree 32 are potentially the largest and, if the small terms of degree 34 and 36 are evaluated from GEM T3, and the 30th-degree values from Cosmos 220 rocket are accepted, Cosmos 185 rocket gives values for degree 32. They are:

$$10^9 \bar{C}_{32,29} = 1.8 \pm 0.9 \quad \text{and} \quad 10^9 \bar{S}_{32,29} = -3.6 \pm 1.4.$$

The corresponding values from GEM T3 are 3.2 ± 1.3 and 3.4 ± 1.3.

So, despite the difficulties in analysing two-day resonances, it seems that the resonance values of order 29 and degree 30 and 32 are competitive with GEM T3 and may prove to be better.

Progress with 27:2 resonance was even slower than with 29:2, because no suitable satellites presented themselves. The first analysis was not until 1988, when Alan Winterbottom determined 90 orbits of Aureole 2 rocket from 7400 observations, including 344 Hewitt camera observations, between September 1983 and December 1984. Three pairs of lumped harmonics of order 27 were evaluated, with an accuracy equivalent to 5 cm in geoid height, which was as good as could be expected in view of the fairly severe drag, but not good enough to improve on the values now available from GEM T3.

Farewell to resonance

The resonance analyses of the 1980s led to what are probably the best values of geopotential coefficients of order 15 (Table 6.3), and to values for orders 14, 16, 28, 29, 30 and 31 which are still competitive with – and may prove to be better than – the values from the best comprehensive models of the early 1990s. Such extensive results were well beyond the fledgling hopes of the early 1970s.

The studies of resonance have yielded much useful information with relatively little effort, and have been valuable in providing independent tests of the comprehensive gravity field models, into which vastly greater efforts have been directed in the 1980s because of the urgent need for better models. This need arises mainly because altimeters aboard satellites can measure the distance between the satellite and the ocean beneath correct to

10 cm. To interpret such measurements properly, both the orbit and the average ocean surface (the geoid) also ideally need to be accurate to 10 cm. In fact the geoid accuracy was about 2 m in the early 1980s and the orbital accuracy similar. Early in 1982 I was invited to a special meeting at Goddard Space Flight Center to discuss how to improve the models: as NASA paid my expenses, I was able to go. As a result of the meeting, more efforts were made to improve the Goddard Earth Models, and GEM T3 is the latest in the line. It is hoped that, with GEM T3, the radial error in the orbit of an oceanographic-type satellite of very low drag may be as little as 20 cm, and that the average error in geoid height may be as little as 40 cm. Presumably the goal of 10 cm for both will eventually be attained; but not easily. (Anything less exact is marginal for oceanographic research, because the height of mid-ocean tides is generally only about 50 cm, distortions due to winds and ocean currents being at about the 10 cm level.) The accuracy of the resonance results is of course better than 10 cm: the 15 pairs of coefficients of order 15 and degree 15, 16, ..., 29 have an average accuracy equivalent to 0.8 cm, and the overall error for order 15 is therefore probably about 2 cm. Larger errors arise in the comprehensive models for other orders (for example, orders 5–10), where no resonance results are available.

The last conference at which I spoke about the resonance results was the COSPAR meeting at Toulouse in 1986. Here the '1987' solution for 15th-order harmonics (Table 6.2) and the solution for 16th-order harmonics (Table 6.6) were first presented.

Winds in the upper atmosphere

From the staid and steady gravitational harmonics, which are not likely to have changed much in the past century, it is quite a leap up to the winds of the high atmosphere, blowing hither and thither in response to transient pressure changes. The techniques of the 1970s continued in use and were applied to orbits newly determined between 1980 and 1982 to give a number of further values of Λ, or of wind speeds at specific local times.

By 1980 there were already good results from three fresh orbit determinations. Analysis of Doreen Walker's determination of 100 orbits of Cosmos 462 from 8600 observations[26] revealed a summer/winter division among the four values of Λ obtained, for heights of 220–250 km. Harry Hiller's analysis of 127 orbits of China 2 rocket determined from 8700 observations,[27] gave three values of Λ averaged in local time and one morning value, for heights of 270–310 km. During the last 12 days, the orbit indicated a west-to-east wind of 240 ± 40 m/s at local time 19–23 h and height 195 km – the strongest well-determined wind derived from orbit analysis. The third satellite was Skylab 1 rocket, for which I determined orbits at 62 epochs from 5000 observations. This satellite, unlike the other two, was in a nearly circular orbit, and the height decreased greatly, from about 400 km initially to 200 km in the last four days. Would Skylab 1 rocket confirm the previous finding that the rotation rate was greatest at heights near 300 km? Yes, it did: the values of Λ obtained were 1.04 ± 0.05 at 380 km height, 1.34 ± 0.09 at 305 km, and 1.06 ± 0.06 at 200 km.

The next two years yielded a further good harvest of values from five analyses of orbits determined from observations. The orbit of China 6 rocket, determined at 51 epochs from 4000 observations, gave eight values of Λ, including three for morning (0.9, 0.7 and 0.8), and two for evening, both 1.2. Daily orbits of Cosmos 1009 rocket during its last 15 days in orbit, when the perigee height was decreasing from 150 to 125 km, gave two low-altitude values of Λ. The orbit of Skylab 1 in its last 15 days, determined from 2000 NORAD observations, gave $\Lambda = 1.10 \pm 0.07$ for a height near 210 km. Two further values came from the orbit of Cosmos 482 in its last 15 days. Lastly, there was NOAA B: orbits at 40 epochs determined from 3000 observations gave $\Lambda = 1.10$ at 300 km for average local time and $\Lambda = 1.15$ at 225 km in the evening.

Armed with these new results, we revised the general picture of the variation of Λ with height and local time – and now season – in a paper[28] published in 1983, which gives the detailed references for the results mentioned in this and the preceding paragraph. After all the data had been examined, 85 values of Λ were accepted as of adequate accuracy and reliability. These included ten from analyses at the University of Aston – of Cosmos 408 rocket by Clive Brookes, of Ariel 1 and Heos 2 rocket by Philip Moore and David Holland, and of Explorer 24 by Graham Swinerd. There were also 14 values from other sources, including five apparently anomalous results by J. M. Forbes, which we were able to recalculate to give satisfactory values. The 85 values were plotted against height, and nine curves (some very short because of the lack of data) were drawn through the points, for morning, evening and average local time, and for winter, summer and average season. (The diagram is not shown as it is similar to the 1988 version, given later.) We summarized the conclusions as follows:

The value of Λ (in rev/day), averaged over both local time and season, increases from 1.0 at 125 km to 1.22 at 325 km and then decreases to 1.0 at 430 km and 0.82

at 600 km. The value of Λ is higher in the evening (18–24 h), with a maximum value (near 1.4) corresponding to a west-to-east wind of 150 m/s at heights near 300 km. The value of Λ is lower in the morning (06–12 h), with east-to-west winds of order 50 m/s at heights of 200–400 km. There is also a consistent seasonal variation, the values of Λ being on average 0.15 higher in winter and 0.1 lower in summer than the average seasonal value.

No significant variation with solar activity was found, but there was a slight tendency for the rotation rate to be greater at lower latitudes, for heights above 300 km. Also the values for the 1970s tended to be about 0.1 lower than in the 1960s, a curious result, for which no explanation has been found.

By 1980 the theoretical models of upper-atmosphere winds were becoming quite realistic, the global model of Fuller-Rowell and Rees[29] being particularly good. This model agreed quite well with our results. At a height of 250 km our average summer Λ was 0.94, with a daily oscillation of amplitude 0.22, while the corresponding figures for the model were 0.95 and 0.23. Our average winter value was 1.20, with daily amplitude 0.26, as compared with 1.12 and 0.20 in the model. For average season our values were 1.13 and 0.25, as compared with 1.04 and 0.18 in the model. Some scatter is inevitable among the results from orbit analysis, because few of the orbits are purely summer or winter, and few are purely evening or morning: most are hybrid to some extent. Also there are the effects of geomagnetic storms (which can reverse the wind direction), and the division between the 1960s and 1970s. So, even if the orbits were perfectly accurate, the possible spread in the curves would be about 0.05. With less than perfect orbits, the curves would be expected to have somewhat larger uncertainties – to call them 'errors' would be libellous. Thus the agreement between the experimental results and the theoretical model at 250 km is quite satisfactory. However, the model does not show the decrease of Λ with height above 350 km.

Soon after this 1983 review was published, Alan Winterbottom and I determined the orbit of Cosmos 482 at 77 epochs from 5850 optical and radar observations.[30] During the $3\frac{1}{2}$ years covered by the orbit determination, the orbital period decreased from 157 to 94 minutes and the inclination decreased from 52.14° to 51.95°, as shown in Fig. 6.11. This satellite was almost ideal for determining atmospheric rotation rates, and eight good values of Λ were obtained, for heights decreasing from 250 to 195 km. The first four values of Λ, all averaged in local time, were 1.05, 1.15, 1.20 and 1.15 with s.d. between 0.02 and 0.05. Of the other four values, two were morning (0.70 ± 0.03 and 0.73 ± 0.03), one was evening ($1.25 \pm$

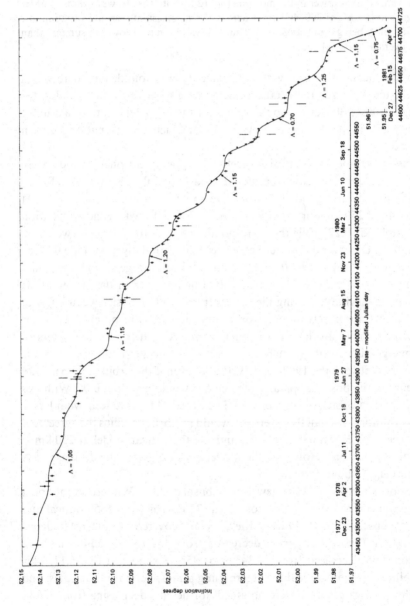

Fig. 6.11. Values of inclination[30] for Cosmos 482, with theoretical curves for various atmospheric rotation rates Λ.

0.04), and one was average (1.15 ± 0.07). All were average in season. These values, the best so far derived from any single satellite, gave strong support to the existing picture of the evening–morning division.

Good results were also obtained at this time from three further orbit determinations at the University of Aston. Two values of Λ came from the orbit of Intercosmos 13 rocket determined at 103 epochs by Brookes and Swinerd from 7000 observations. Seven further values of Λ were derived from two extensive orbit determinations by Boulton, of Nimbus 1 rocket at 285 epochs from 14000 observations, and of Intercosmos 10 rocket at 296 epochs from 14000 observations – the lengthiest orbit analysis yet completed. (References for these papers are given in the review discussed below.)

In 1987 we added these new values, to update the previous review. There were no significant changes in the picture with the new total of 103 values,[31] though some changes in the positions and shapes of the curves were necessary. The 85 values that are seasonally average are shown in Fig. 6.12, with the morning, evening and average curves as before. The results were summarized as follows:

The average value of Λ increases from 1.0 rev/day at 125 km height to about 1.2 at 300 km, and then decreases to about 0.9 at 500 km; Λ is higher in the evening (18–24 h), with a maximum value near 300 km which corresponds to a west-to-east wind of 150 m/s, and is lower in the morning (06–12 h), with east-to-west winds of order 50 m/s at heights of 200–400 km.

Again the rotation rate was greater in winter than in summer, for heights of 200–300 km. A possible variation with solar activity was also apparent, Λ being about 0.1 lower for medium solar activity than for high or low solar activity.

During the 1980s various new methods were developed for measuring upper-atmosphere winds either from measurements *in situ* by instruments in spacecraft or from ground observations. These new techniques offer more detailed and localized information than orbit analysis can provide. So, in one sense, the evaluation of Λ by orbit analysis could be called outmoded. However, the localized measurements do not give a world-wide average Λ, and the values derived from orbit analysis will probably continue to be used on the many occasions when it is necessary to remove the effects of atmospheric rotation on an orbit, for example when analysing resonances.

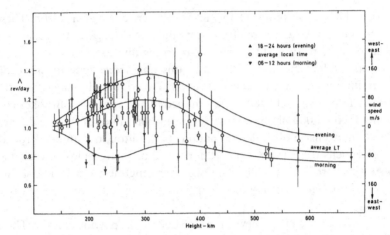

Fig. 6.12. Variation of atmospheric rotation rate Λ with height and local time, for
average season, as derived[31] in 1988.

Researches on upper-atmosphere density and structure

Our research on the upper atmosphere took a different form in the 1980s.
At the RAE there were no lengthy studies of the variation of density over
several years, to determine the semi-annual variation. Instead the researches were based on detailed and accurate orbit determinations from
observations of a limited number of satellites that suffered fairly severe
drag, often those in the last two weeks of life. The long-term studies of
density continued at the University of Aston, where several excellent
records of the variations of air density over the years were derived.

Values of density scale height *H* were obtained from all the high-drag
orbits determined from observations. The results were more accurate than
in the 1970s; so departures from the *CIRA* model could be pinned down
unambiguously. The accuracy was particularly good for the daily orbits in
the last two weeks of the satellite's life derived from the specially-requested
NORAD observations. This was because the perigee height always
decreased greatly during these weeks. For example, Harry Hiller's orbit
determination of Cosmos 1009 rocket in its last 14 days[32] gave five good
values of *H*, of which the first three agreed with *CIRA*; the fourth was
larger, but at the time of a magnetic storm. The fifth value, for the last day in
orbit, was also unusually large; but the perigee height was decreasing from
135 to 126 km, a height where the mean free path of the air molecules is
7 m. As the satellite was 8 m long, free-molecule flow would be changing to
transition flow during the day, thus reducing the drag. It was the 'drag
scale height' rather than the density scale height that was being measured,

and the former would be larger. Hiller also determined the orbit of Heos 2 rocket in its last 16 days,[33] obtaining eleven values of H, each accurate to about 2 %: they were about 4 % higher than given by *CIRA* . So far, *CIRA* was generally being substantiated.

A new trend emerged when orbits began to be determined for the years 1980–81, when the solar activity was higher than at any time since 1958, and therefore higher than at the time of the (mainly 1960s) observational values on which *CIRA* was based. The orbit determination of NOAA B for 1980–81 gave ten values of H, each accurate to about 3 % for heights of 280–370 km, and they exceeded the *CIRA* values by about 15 %. We concluded that the *CIRA* values of H (and presumably also density) were too small when solar activity was high. This was confirmed by another orbit determination, of Cosmos 482 in its last 15 days in orbit in May 1981. The values of H, again accurate to 3 %, were 6 % greater than given by *CIRA*, the difference being less here because the height was much lower (180–200 km). All this added a mite to knowledge, but was not the red-hot 'exciting science' that appealed to Research Councils struggling in straitened circumstances.

More interesting, though possibly still not high on the excitement scale of the Research Councils, were the other varied features of the upper atmosphere investigated with the aid of orbit analysis in the 1980s. The first new departure was analysis of the longitude of the node Ω to determine zonal and meridional winds. The possibility of this procedure had been known since 1959, but the first successful application was with NOAA B in the last 8 days of its life.[34] The variation of inclination over these days gave $\Lambda = 1.15$, and, since the satellite was in a very nearly polar orbit, the gravitational perturbations of Ω could be subtracted out quite accurately. Analysis of the remaining variation of Ω confirmed the value of 1.15. A rather similar analysis, but with a different twist, was possible for Cosmos 482 in the last few days of its life, when Ω was affected by both zonal and meridional winds.[35] On removing the gravitational effects and those of atmospheric rotation (with the value of Λ found by analysing inclination), the remaining change in Ω revealed a meridional wind (south–north) of 20 ± 20 m/s in April 1981 at latitudes 30–50° S at local time centred on 08 hours. Though scarcely riveting, the results show the capabilities of the technique.

Two other structural features of the upper atmosphere were measured by analysing the orbits of Cosmos 1009 rocket and Cosmos 482. For the former in the last 4 days of its life, the argument of perigee ω increased by 4° more than it would have done in a spherically symmetrical atmosphere.

This variation in ω was caused by atmospheric oblateness, and yielded two values for the ellipticity ε' of the atmosphere: it was equal to the Earth's ellipticity ε ($= 0.00335$) in the first two days – which was no surprise – but decreased to $\frac{1}{2}\varepsilon$ in the last two days (3–5 June 1978). Presumably the change was linked with a magnetic storm on 2 June, which may have increased the density at high latitudes. For Cosmos 482 in its last 15 days, the analysis of ω went one stage further, because the perigee height was greater (180 km initially) and an estimate of the day-to-night variation in density was also possible. Fig. 6.13 shows the values of ω after removal of gravitational effects, with various theoretical curves. For a spherically symmetrical atmosphere, the change in ω is very small and quite unlike the real variation. For oblate atmosphere (with ellipticity 0.00335), the dot–dash curve, the agreement is better. When the effect of the day-to-night variation is added, the agreement is much improved, being best if the strength of the day-to-night change is taken as 0.7 of that given in *CIRA*.

The orbit determination and analysis of Skylab 1 in its last 14 days[18] was the most productive of all these analyses of individual satellites. As well as yielding values of 16th-order harmonics and atmospheric rotation, Skylab 1 also gave accurate values of air density at heights from 252 km down to 179 km from 28 June to 11 July 1979. This was possible because the mass/area ratio was known, and the results provided a good test of *CIRA*. In the event the values of density from Skylab 1 were within 1% of the *CIRA* values for 28 June–6 July, and within 4% for 7–11 July. In view of the uncertainty of perhaps 5% in the drag coefficient, this can be called perfect agreement.

This was a worthy but rather dull result: more interesting – especially to those who had developed the theory – was the chance of seeing exactly what happened to the eccentricity of this near-circular orbit as decay approached. According to the theory described in Chapter 4, a near-circular orbit under severe drag in an atmosphere with day-to-night variation in air density should wriggle around in such a way that the perigee moves towards the point of lowest density on the orbit. But the gravity field is also trying to move the perigee round a circle in the (ξ, η) plane (Fig. 4.9). Presumably a close look would reveal drag gradually overcoming gravity as decay approached. The best previous illustration of this process was with Skylab 1 rocket, where the motion of perigee during the last 11 days could be seen to be governed by the known force produced by the gravity field and an increasing atmospheric effect. However, this satellite was of higher drag and did not have so nearly circular an orbit as Skylab 1 itself.

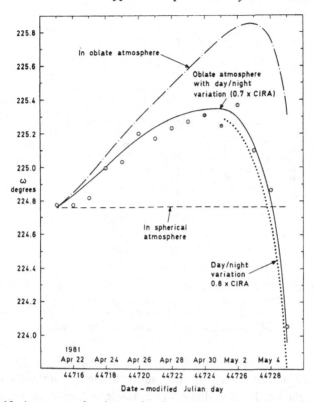

Fig. 6.13. Argument of perigee ω for Cosmos 482 during its last 15 days in orbit, after removal of gravitational perturbations, with theoretical curves for comparison.[35]

Fig. 6.14 shows the variation of eccentricity in the last 15 days of the life of Skylab 1. The quantities plotted are $z\cos\omega$ and $z\sin\omega$, where $z = ae/H$ as usual. The initial value of eccentricity e was 0.000907, so that the perigee height was only 6 km lower than the apogee height: the error in this and all the values of e was about 5×10^{-6}, equivalent to 30 m. The initial value of ω was 102° and, as shown in Fig. 6.14, this gives an initial value of $z\sin\omega$ of about 0.13. In the absence of drag, the track of z in Fig. 6.14 would be an anticlockwise circle centred on a point very close to the $z\sin\omega$ axis at a distance of about 0.12 from the origin (though this distance increases slightly as time goes on and H decreases). The first four observational values in Fig. 6.14 are close to the gravitational circle, but after that the locus diverges.

The rationale for the actual track can be seen from the triangles, joined with a broken line, in the lower part of the diagram. These triangles are the

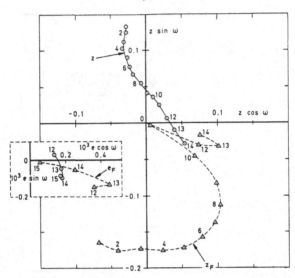

Fig. 6.14. Evolution of the eccentricity e (or $z = ae/H$) for Skylab 1 during its last 15 days in orbit.[18] The successive daily orbits are numbered 1–15. The circles give the observational values, the triangles the theoretical final values z_F or e_F that would be appropriate if drag were all-powerful.

end-values of z predicted by the theory if the atmospheric drag were all-powerful and did not have to contend with the gravitational forces produced by the Earth's pear shape. For example, at orbit 6, the satellite is being pulled towards the triangle marked 6 by drag and pulled to the right (as seen on the diagram) by the gravitational forces, but less strongly. The whole locus can be looked at in this light, and the results in the last three days, orbits 12–15, are shown separately in a smaller diagram in terms of e rather than z. At July 11.0 on orbit 14, e is still increasing, but it decreases on the final orbit 15, for July 11.5, in response to the very small 'end-value' of e at this epoch. The orbit determinations of Skylab on its last day in orbit were particularly accurate because of the large number of observations (about 800) available between midday on 10 July and the decay at 16.30 UT on 11 July.

A renaissance of theory

The main theory for the effect of air drag on orbits had, it seemed, gone to sleep in the 1970s, after the rapid appearance of Parts I–VI between 1960 and 1968. But, like the Sleeping Beauty, the theory was only asleep, not dead, and was woken again after fourteen years when Bill Boulton and Graham Swinerd at the University of Aston began a systematic extension of the theory for $e < 0.2$, including many smaller terms that had previously been neglected, combining the effects of oblateness and day-to-night variations (instead of just assuming they could be added), and extending the theory to other orbital elements, particularly the mean anomaly. Their six papers in the *Proceedings of the Royal Society*[36, 37] in 1982–84 were most valuable in consolidating the theory by deriving more accurate forms for results over a single revolution.

At the RAE in the 1970s there had been little work on the theory of air drag, apart from systematizing the lifetime calculations. The somnolence continued into the early 1980s because of the unfavourable climate for orbit analysis. But a renaissance began in 1984, when Dr Keith Whittles of the publishers Blackie asked me to produce a new and much-expanded edition of my book on the theory of satellite orbits in an atmosphere. No one else was likely to make such an offer, so I accepted. Several gaps in the theory were waiting to be filled, and with the incentive of the publisher's deadline I began to give the missing theory priority over the analysis of actual orbits. The result was three papers published in the *Proceedings of the Royal Society* in 1987, all in collaboration with Doreen Walker, and a revision of the methods for lifetime estimation.

The first of the three papers[38] was the long-delayed Part VII of the main air-drag theory, in which the high-eccentricity theory of Part III was extended to a more realistic atmosphere with the scale height H varying linearly with height, as in Part IV. The rate of change of H with height, μ, was again assumed to be less than 0.2. The fruits of this marriage of Parts III and IV proved to be quite useful. It turns out that the constant-H theory for the variation of perigee height with eccentricity – the equation on p. 99 – can still be used if H is evaluated at a height $1.4 H_{p0}$ above the initial perigee height y_{p0} (or $1.5 H_p$ above perigee y_p for single-revolution applications). The effect of μ on the variation of eccentricity with time is very slight, and the variation of z ($= ae/H_p$) with time remains very nearly linear. Also the constant-H formula for lifetime in terms of the decay rate $\dot{T}$ still applies, with error of order $0.5\mu^2$. None of these results was startling, but they were needed to interpret analyses of high-eccentricity orbits, if only to give estimates of the errors incurred by the use of the previous over-simple theory.

The second paper was a new approach to the theory for the effects of meridional (south-to-north) winds on orbits, largely superseding the previous theory of 1966. The previous theory covered orbits of eccentricity up to 0.2, but required some rather savage approximations. In the new paper[39] a different controlling parameter was introduced, leading to a much simpler and more accurate theory for any eccentricity between 0 and 1; the effects of atmospheric oblateness were also covered. The new procedure was to take as a measure of the south-to-north wind, V_{SN} say, the quantity M, where $rM = V_{SN} \sec \phi$. (Here ϕ is the latitude and r the distance from the Earth's centre, as usual.) The use of M might seem dangerous, as it would become infinite at the poles; but meridional winds have no sideways effects on exactly polar orbits, so the danger proves to be illusory. The theory was developed, with M constant, in two parts, for e less than or greater than 0.05.

The theory shows that the change Δi in inclination i caused by a meridional wind M is related to the change ΔT_d in orbital period (expressed as a fraction of a day) by the equation

$$\Delta i = -\Delta T_d \, M' \, \alpha' \cos i \cos \omega,$$

where M' is the value of M expressed in revolutions per day, and α' is a factor dependent on e: the value of α' increases from 0 at $e = 0$ to 0.25 at $e = 0.04$, and then decreases steadily to 0.05 at $e = 0.4$. In practice the meridional wind, and hence the value of M, is likely to vary with latitude: for high-eccentricity orbits this does not matter, because the effects are concentrated in a small range of latitudes near perigee; for eccentricities less than about 0.01, the assumption of constant M is not so realistic because the effects are spread in latitude. Despite these practical limitations, the theory nicely encompasses the effect of meridional winds on all orbits.

The third paper, Part VIII of the series, tackled a thorny and long-postponed problem, the theory for orbital lifetime in an oblate atmosphere when the perigee distance oscillates in response to the odd zonal harmonics.[40] Until now, empirical corrections had been devised to assess these effects. Part VIII provided the necessary theory for calculating two multiplying factors, written as $F(oz)$ and $F(ao)$, which express the effects of odd zonal harmonics and atmospheric oblateness on the lifetime as calculated in the absence of these effects. The results are valid for all eccentricities between 0 and 1, and cover nearly all possible orbits. Both correction factors depend on the current and final positions of the argument of perigee ω, written as ω_0 and ω_L respectively. The variation of $F(oz)$ with ω_0 for a series of values of $\omega_L - \omega_0$ is shown in Fig. 6.15; the heavy line applies if $\omega_L - \omega_0$ is equal to 1, 2, 3, ... revolutions, and is a useful average. The results are for $b = 0.2$, where $b = (aK/H)(1 - 1/2z_0)$ if $z (= ae/H)$ is greater than 0.5. Here aK is the amplitude of the variation due to odd zonal harmonics, and Fig. 6.5 shows how aK depends on

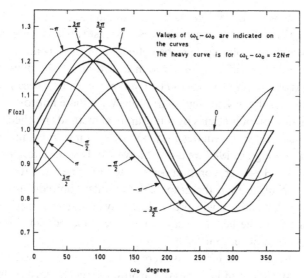

Fig. 6.15. The factor $F(oz)$ by which the lifetime should be multiplied to allow for the effects of odd zonal harmonics:[40] ω_0 and ω_L are the initial and final values of ω. The curves apply for $b = 0.2$, where b depends on aK (Fig. 6.5), as explained in the text (and more fully in the original paper[40]).

inclination. Usually b will differ from 0.2 and, as b controls the amount by which $F(oz)$ departs from 1.0, the value of $F(oz)$ in Fig. 6.15 should be correspondingly scaled.† The graph for $F(ao)$ has a form similar to Fig. 6.15, but with the scale for ω_0 running from 0° to 180° and the maximum value of $F(ao)$ being near $\omega_0 = 0°$ (or 180°). Either atmospheric oblateness or odd zonal harmonics can cause changes of up to 30% in lifetime; so their effects need to be carefully assessed. However, the effects are not always so large, and the factors can be near 1.0 for some values of ω_0 and ω_L, as Fig. 6.15 shows. Both effects are small for inclinations less than 30° and for near-circular orbits.

With the results of Part VIII available, it was time to revise the manual of satellite lifetime estimation. This was done in 1987, and the 'bible' appeared in full as RAE Technical Report 87030, with a shortened version published in a journal.[41] The procedure was to calculate a 'basic' lifetime $L^* = Q/\dot{n}$ (with Q as in Fig. 5.10) and then to take the true lifetime L as

$$L = L^* \times F(oz) \times F(ao) \times F(sa) \times F(sc) \times F(dn).$$

The three extra factors here take account of the effects of the *semi-annual* variation in density, the variation of density during the *solar cycle* and the

† Explicitly, $F(oz) = 1 + b\{\sin \omega_0 + (\cos \omega_L - \cos \omega_0)/(\omega_L - \omega_0)\} + O(\tfrac{1}{2}b^2)$.

*d*ay-to-*n*ight variation in density. The semi-annual factor had already been in use since 1978. The solar-cycle factor is the least satisfactory because it requires a 'standard' solar cycle to be defined, and there is no such thing – every solar cycle is different in strength and timing. The day-to-night factor $F(dn)$ is of a similar format to $F(sc)$, but more realistic because of the basic regularity of the day-to-night variation. With these improvements in 1987, the calculation of lifetimes by our graphical methods reached its final form. In future, lifetime estimates may be made by numerical integration on computers, though in 1987 we did not have a fully satisfactory program for this purpose. However, computer-generated lifetime estimates will not be any more accurate than the graphical methods. Unpredictable future variations in solar activity will probably always cause errors of 10 % or more, and if this is the best that can usually be achieved, small errors in reading graphs are not important.

Although funding organizations were not always impressed by orbit analysis, the estimation of lifetimes did arouse interest because of hype by the media about the 'danger' posed by the decay of selected satellites. As in the 1970s, two large satellites per week were descending and scattering fragments on the sea or ground; so the scaremongering could arise at any time. No one has yet been injured by a falling satellite, while many people on the ground have been killed or injured by aircraft, of which more than a hundred per week fall out of the sky world-wide, and preferentially over populated areas. Though we were sceptical about the danger, any interest in the orbit work was welcome in the early 1980s.

A new 're-entry crisis' arose at the end of 1982 after the malfunction of Cosmos 1402, another satellite with a nuclear reactor as an on-board power-source. The reactor and its fuel core failed to go into the intended high orbit, but were successfully separated from each other. The reactor decayed on 23 January 1983 and the core on 7 February, the latter probably burning up completely. We were involved in predicting the descent dates, and did so with an accuracy better than the 10 % expected.

In September 1985 the European Space Agency organized a meeting on 'Re-entry of space debris' at Darmstadt, where our current methods of lifetime prediction were presented. The same subject was covered in 1987 when the finalized method was presented as my last conference-paper, at the 38th International Astronautical Congress at Brighton in October. This was a memorable occasion, because it was combined with meetings, and a reception in the Royal Pavilion, for members of the International Academy of Astronautics, and above all because it ended with the 'hurricane' of 16 October 1987, which struck with maximum force at

Brighton, bending steel street signs, uprooting nearly all the trees near the seafront, and smashing in the front of the conference centre. So my scientific conference-going ended with the lower atmosphere trying to emulate the wind speeds we had so often found in the upper atmosphere, blowing me back into the real world with éclat.

Back into the outer world

I mentioned earlier in this chapter the creation of the UK Facility for the Precise Tracking of Satellites in 1984. It seemed very promising, and for three years all went well: good work was done, in observations by the laser and cameras and in their analysis for geophysical research. But the Facility was not allowed to run freely any longer. In 1987 the Astronomy, Space and Radio Board of the SERC, itself facing financial cuts, decided on a 'review'. The chairman of the review panel – would the 'chair' be transmogrified into 'axe', everyone wondered? – was Professor J. R. L. Culhane, and I was the MoD representative. We had three amicable meetings in the summer of 1987, with no hint of a bad outcome. Then, at the end of the third meeting, some 'conclusions' pretyped by the SERC were produced by the secretary. They were quite unconnected with the previous discussions and recommended that the SERC should cease its involvement with the Hewitt cameras, while keeping the laser. The other members of the panel, the proverbial 'impartial outsiders', supported my protest that the meetings had been a farce, as these conclusions had never been mentioned. So a compromise was reached, that the SERC should 'discuss the matter with its partners', while I pointed out to the MoD that the RAE only used the Hewitt camera observations and not those of the laser; to close down the cameras would therefore be self-destruction. A strong hidden force was lurking behind, however: the SERC was in process of slimming the RGO, before removing the residual staff to Cambridge, a move to be financed by selling off Herstmonceux Castle. So it was no surprise when the 'discussions with partners' led to the decision that the Hewitt cameras, after a two-year stay of execution, should be closed down in March 1990, the date when the remaining RGO staff were scheduled to be transferred to Cambridge.

This made 1987 quite depressing, and a further problem arose in October, when the Vice-Chancellor of Aston University refused to accept the grant of £120000 made to the University by the UK Facility for the Precise Tracking of Satellites. As a result of this refusal, the RGO agreed to take over the cameras, the prediction service – and the grant – from

Aston in March 1988. That was the none-too-promising situation when I retired from the RAE on 3 May 1988 – for I was to my surprise given the privilege of an extra six months' employment after my 60th birthday.

Though the cameras and prediction service suffered trauma and transplantation, the Optical Tracking Subcommittee continued unshaken through the 1980s, as mentioned earlier. As chairman, I reported on the meetings to its parent, the British National Committee on Space Research, of which Sir Harrie Massey was chairman from 1959 to 1982, and Robert Wilson from 1983 to 1989 – after which all the National Committees were 'dissolved'. I had been appointed to the National Committee in 1965 and attended every meeting from then until its dissolution. From 1978 onwards I was Deputy Chairman: this was normally a sinecure, because Sir Harrie Massey (and later Robert Wilson) took all decisions as chairman; however, I sometimes had to act on urgent matters when Sir Harrie was away on one of his frequent overseas visits. The Committee was notable for the many Royal Society discussion meetings that it promoted. It was also the UK link with COSPAR. At the COSPAR meeting in Ottawa in 1982, Sir Harrie Massey and I were the UK delegates; apart from this, my only visit to a COSPAR meeting in the 1980s was to that at Toulouse in 1986.

There were several organizational changes in Britain in the second half of the 1980s; some seemed promising, others depressing. The death of Sir Harrie Massey in November 1983 left British space research without a captain. His had been a wise, immensely knowledgeable and impartial voice, listened to with respect by university and government committees alike. He was irreplaceable, because the other leading university researchers, all competing for funds that became scarcer every year, could not achieve credibility as impartial advisers.

As if in recognition of the yawning gap, the British National Space Centre (BNSC) (first discussed in 1958) was set up in 1985. The Director was Roy Gibson, previously the first Director-General of the European Space Agency, and everyone welcomed this new initiative, which held the promise of putting space policy 'on the rails' again. Roy Gibson's energetic and persuasive advocacy of space enterprises won golden opinions from everyone – except the Government, which declined to fund the programme. Consequently, Gibson resigned in 1987 and since then the BNSC has remained a low-key organization, usefully channelling the Government funds allocated, but not taking independent initiatives. The euphoria of 1985–86 was short-lived.

At the RAE, cuts and reorganizations continued through the 1980s, and Space Department was split in 1986. Doreen Walker and Alan Winterbottom, accompanied by 57 filing cabinets of satellite data, were transferred to another building 400 m away (R14) under the aegis of a new department called Special Systems. I was also transferred to the new department but remained in Q134 Building because I was due to retire so soon. The disruption was bad enough, but the cutting of twenty-five years of links with people in Space Department was even worse.

Throughout most of the 1980s the research fortunately had an intrinsic vitality that overcame all the obstacles created by the organizational problems mentioned, and many unmentioned. At the University of Aston, Graham Swinerd and Bill Boulton, free of lecturing commitments, were most productive, with numerous papers in *Planetary and Space Science* to add to the six in the *Proceedings of the Royal Society*. Philip Moore continued with his innovative work in orbit analysis and began using the laser observations. Clive Brookes himself also contributed, though increasingly inhibited by administrative problems as the decade progressed, culminating in the refusal of the grant by Aston in 1987, which was virtually the end of ESRU. The work at the University of Leicester had declined by 1984 with the departure of Stefan Hughes, though Jack Meadows continued to be most helpful in the organizational battles. Even at the RAE the orbit analysis continued to flourish, as is evident throughout this chapter. And, though the RAE was shy, there was some recognition from outside. Doreen Walker was elected a Member of the International Academy of Astronautics in 1985, and I received an honorary doctorate from the University of Surrey in 1986.

In the early 1980s, partly impelled by the problems over the support of orbit analysis, my interests turned more towards the history of science. My edition of the *Letters of Erasmus Darwin* was published by Cambridge University Press in 1981, after I had spent several years in searching, editing and identifying unknown names. By good luck the book came out just before the 250th anniversary of Darwin's birth (on 12 December 1731), and I was able to 'launch' it with a talk at Darwin College, Cambridge, followed a few weeks later by the Sydenham Lecture of the Worshipful Society of Apothecaries, when I again spoke about Erasmus Darwin. In 1982 I was asked to be one of the H. L. Welsh Lecturers at the University of Toronto in May: Erasmus Darwin, satellite observing and orbital theory were the topics of the three Toronto lectures. For the Milne Lecture at Oxford in 1984 I also took a historical theme, the history of ideas on the Earth's atmosphere.[42]

Another strand of interest in the history of science was via the Royal Society and its National Committee for the History of Science, Medicine and Technology, to which I was appointed in 1979. The Committee was responsible for awarding grants in the history of science, for the general welfare of the subject in Britain, and for international aspects. I became chairman in the autumn of 1984 and have subsequently been much involved in the administrative aspects of the subject though (regrettably) not in the scholarly detail.

The most pleasing event I initiated as chairman of the Committee was an international meeting at the Royal Society on 30 June 1987 to celebrate the 300th anniversary of the publication early in July 1687 of Sir Isaac Newton's *Principia*, perhaps the most famous of all scientific books. Most of the organizing was done by my oldest friend in the history of science, Professor Rupert Hall, a leading authority on Newton. My opening remarks to the meeting stressed my own Newtonianism:

Although I am here as Chairman only because I am Chairman of the National Committee, it so happens that I am not just a cog in the machinery but one of the most persistent of modern Newtonians. As a teenager I went to Trinity College Cambridge because of Newton, and since then I have spent 30 years extending the theory of the *Principia* to the orbits of Earth satellites. I have never heard of Einstein; for me, Newton's laws still rule.[43]

This was a good note on which to proceed to the scholarly papers that followed. A few days before the meeting, Newton exhibits had formed the centre-piece of the Royal Society's summer Soirée, and we provided one of these exhibits, showing Newton's treatment of Earth satellites and the use of orbit analysis to determine the shape of the Earth.

A few weeks later there was another celebratory conference on 'Celestial Mechanics 300 years after the *Principia*', held at the Royal Greenwich Observatory, Herstmonceux, from 20 to 24 July. Our contribution,[44] entitled 'The effect of air drag on satellite orbits: advances in 1687 and 1987', interpreted in modern terms Newton's treatment of the effect of air drag on satellites, compared his and the modern theories, and then (as rather an anticlimax) went over the thin sheaf of papers published in 1987. This was another memorable conference, unrepeatable in several respects: it was the last major RGO conference to be held at Herstmonceux Castle; the conference dinner at Drusilla's near Alfriston, the most entertaining I experienced in all my scientific conference-going, was followed by the heaviest downpour of rain that I have known in Britain. Though coincidences do not usually impress me, it does seem strange that my last two scientific conferences both ended with climatic extremes; that their venues were so close, Herstmonceux being only 20 miles from Brighton; and that my birthplace, Seaford, is half-way between them.

In the difficult years of 1981–83 I had begun work on a new book that was published by Macmillan early in 1986, *Erasmus Darwin and the Romantic Poets*. It was good escapism to read through the works of all the poets of the Romantic era to see how much they borrowed from Darwin. The answer was 'a surprising amount', and this is my favourite among the books I have written because it shows so clearly what the literary pundits had failed to notice – the verbal resemblances to Darwin in the work of Wordsworth, Coleridge, Shelley, Keats and many poets less well known. The book had some good reviews, but has so far been virtually ignored by the literary fraternity, thus confirming my low opinion of their intellectual efficiency. Much more successful was a new version of *Observing Earth Satellites* published by Macmillan in 1983; also *Shelley: his Thought and Work* came out in a third edition in 1984. A second book of poems, *Animal Spirits*, emerged in 1983.

Early in 1985 I returned to orbital theory and worked hard for two years on the book that was published in September 1987 as *Satellite Orbits in an Atmosphere: Theory and Applications*. The publishers, Blackie, produced a good-looking book, which was very satisfying for me because it recorded in concise and final form the theory for the effect of air drag on orbits that had occupied a fair proportion of my thirty years in space. But again I was appalled by the number of new misprints that appeared after the final proofs had been corrected: some were linked with the corrections; some were simply corrections not made; some were quite unnerving, such as whole lines disappearing. (I have a list....)

At the RAE the 1980s seemed a gradual run downhill, with few highlights to remember. The abiding disaster of Britain in the 1980s was the continual disintegration of cultures as knowledgeable people melted away. It happened everywhere, merely by default, and nowhere more obviously than at the RAE. When scientists full of knowledge retired – and they were often lured into retiring early – the pressure on numbers of staff ensured that they were rarely replaced. At worst, their speciality was ended; at best, a younger colleague would carry it on with reduced staff and status. New recruits still came in, but they tended to fill the 'high-priority' posts, meeting operational and military requirements rather than pursuing science. So the scientific knowledge and expertise for which the RAE used to be famous was gradually eroded in the 1980s. A crucial step was the abolition of the Departmental Libraries, through lack of staff. Space Department suffered badly, because Anne Maxwell, the Space librarian, had kept everyone well informed. It was she who told me in 1974 that 'a Dr Brookes' had been asking for our reports: this was a trigger for the creation of ESRU. If the requests had been impersonal, by computer, I should never have known of them. After the closure of the Space Library on her retirement, there was much greater danger of different people doing the same work unknown to each other.

Even in the 1980s, however, there was much that was admirable at the RAE. As in earlier decades, the tracers in the Drawing Offices produced accurate diagrams that enhanced the appearance and appeal of our papers. The scientific typists in the Printing Department miraculously transformed scruffy manuscripts into authoritative-looking typescripts. These texts and diagrams were brought together into RAE Technical Reports, printed and produced in the Printing Department under the benevolent rule (and eagle eye) of Peter Rolls. My daily dictation sessions also continued, though eventually there was only one shorthand typist at the RAE. How I miss her, and wish I could afford that luxury in retirement!

7

Out of the fray, 1988–1991

Tomorrow to fresh woods, and pastures new.

John Milton, *Lycidas*

Retiring from paid employment at the RAE in May 1988 proved to be the prelude to two years of unpaid attendance part-time, clearing up the loose ends. During the last twenty of my forty years at the RAE I was able to adopt the most efficient procedure of filing away all working papers and reports received, by subject, in filing cabinets that remained in place and just increased in numbers. No time was wasted in going through them on throwing-away sprees. It was 'onward undaunted' continuously, with everything undisturbed, in the same office. All that had to end in 1988. In a long and traumatic series of evening massacres at home, I ploughed through the 25 000 neatly-filed letters in my 'general correspondence' and threw away about 98%. My wife with great forbearance allowed the smallest bedroom in the house to be lined with shelves and converted into an archive room, to store the papers needed for this book, and some of my books and reports on space topics. A second round of massacres is now in prospect among those archives....

At the RAE, meanwhile, I was obliged to vacate my office in Q134 Building in May 1988, and took a suitcase-full of selected papers each day across to a new office in R14 Building, reducing the bulk to a mere six filing cabinets. The Table of satellites was taken over by Doreen Walker and Alan Winterbottom: Geoffrey Perry, uniquely knowledgeable in current space activities, continued to supply the basic data under contract.

During 1989, on the initiative of Peter Rolls, the RAE agreed to print and publish a new edition of the Table of satellites, listing launches up to the end of 1989. Doreen Walker, under contract after her retirement, undertook the task of editing, up-dating and checking in the spring of

217

1990, with help from Alan Winterbottom, and this fourth edition appeared in the autumn of 1990. The book[1] runs to 1056 pages of A4 size, weighs 2 kg, and is easily distinguishable from its plain-clothed predecessors by its fetching pictorial cover-design. The printing was done – well, as ever – by the RAE Printing Department, one of its largest tasks, with over a million pages to be printed. The monthly issues have continued subsequently, with Alan Winterbottom in sole charge.

Retirement made no difference to my satellite observing, always done from home in the evenings, and I continued as chairman of the Optical Tracking Subcommittee. By 1989, however, the mania for cutting that characterized the 1980s had spread to the Royal Society: all its twenty-nine National Committees and their twenty-three Subcommittees were to be 'dissolved' (that means abolished) at the end of 1989 and replaced by one International Relations Committee. I heard about this plan at an early stage and, when the Optical Tracking Subcommittee met a few days later (on 8 March 1989), it sent in a strong plea for clemency, which is worth quoting because it sums up much in this book:

In the past 25 years, about 300 scientific papers have been published giving the results of research into the properties of the upper atmosphere and Earth's gravitational field derived from analysis of satellite orbits determined from observations. A substantial and essential fraction of these observations (essential because of their geographical spread) have been made at minimal cost by volunteer visual observers in the UK and overseas. The programme of observation and choice of satellites for observation have been supervised by the Optical Tracking Subcommittee, and the observing equipment is held at the Royal Society and loaned to observers as recommended by the Subcommittee. The Subcommittee has a unique role in enabling representative volunteer visual observers, who make up the majority of the members, to cooperate with the scientists who use the data. The observation programme would have collapsed many years ago without the Royal Society's support, and will collapse now if support is withdrawn. We therefore request that the Subcommittee be kept in existence after the abolition of the National Committee.

This plea was accepted by the Royal Society, and a new 'Optical Tracking Working Group' was set up in 1990 with the same membership as the Subcommittee. It was the only one of the twenty-three Subcommittees to remain in being.

In March 1990 the two Hewitt cameras closed down, as scheduled, and the four observers had to seek other employment. The Siding Spring camera was taken to pieces and stored in the basement of the Anglo-Australian Telescope building at Siding Spring Observatory. The Herstmonceux camera is still on site as I write, because the future of the Equatorial Group of telescopes there is still uncertain.

Nearly all the remaining staff of the RGO moved to Cambridge in April 1990, but the SERC did accept a recommendation from the scientists concerned that the laser ranger should remain at Herstmonceux, where the site was tied into the world geodetic system. The laser continues to operate there and has become the most productive in the world. The RGO also agreed to continue providing a prediction service for visual observers, as requested by the Optical Tracking Subcommittee in 1989. The predictions have been sent out from the Herstmonceux outstation, and a good service has so far been maintained, including the essential task of recording the observations sent in.

Thus visual observing still continues in fairly good health, though several observers have withdrawn, finding their morale undermined by the apparent lack of appreciation for their work. That is one way of looking at the sorry story of the late 1980s. But a wider view would be to see the closure of the Hewitt cameras as entirely in keeping with the general decline in UK funding in science and technology during the 1980s – from 0.35% of the national GDP in 1981 to 0.28% in 1990, according to the Royal Society. Even though the Hewitt cameras remained to the end the most accurate cameras in the world for satellite tracking, they were seen as outmoded beside laser trackers and electronic imaging methods. If there was not enough money for everything, the sacrificial victims would be the Hewitt cameras, which some 'hitech' scientists found almost embarrassing in their antiquity — superb vintage steam-engines of space research, as it were.

'If the Hewitt cameras are unwanted, visual observations must be supremely useless' was the other morale-sapping thought. But the removal of the cameras from the scene gives visual observation greater importance. To put it succinctly, orbits determined from Navspasur observations alone, all from one latitude, cannot be guaranteed to be free of bias. But Navspasur observations *plus* visual observations from higher latitudes allow the computation of reliable orbits of an accuracy adequate for resonance and other gravity-field studies. So visual observations can still play an essential role, even if no one gallops about proclaiming how wonderful they are.

Of course the observations are of no use if no one uses them, but that situation has not yet arisen. In the years 1988–91 orbit analysis was being pursued at several centres. At the RAE Doreen Walker was completing the determination and analysis of the orbit of Cosmos 184 rocket, one of the best studies of 15th-order resonance so far, and Alan Winterbottom was making the first analysis of 27:2 resonance after determining the orbit of Aureole 2 rocket.[2] Both these were mentioned in Chapter 6. The effort on orbit analysis has subsequently declined, though Bob Gooding has continued his pioneering work on orbital theory, in which he is a world leader. Just before my retirement I collaborated with him in a detailed exposition of the resonance theory, published[3] in 1989.

At the universities there has been some upswing since 1988. In 1987 Graham Swinerd was appointed as a Lecturer in the Department of Aeronautics and Astronautics at the University of Southampton, and, with the help of successive research assistants, has continued to work on orbit analysis, including studies of the semi-annual variation in air density and several resonance analyses, with excellent results on a satellite of very low drag, Meteor 28, at 15th-order resonance. Philip Moore at the University of Aston, the most innovative and productive of the university researchers, has concentrated on the analysis of orbits determined from laser observations, but has also taken on research assistants to pursue orbital studies with optical and radar observations. In particular, Mark Gilthorpe has determined the orbit of Cosmos 1603 and analysed its unusual experience of 14th-order resonance in 1987, when the satellite became trapped in resonance at a time of fairly low solar activity. As well as the usual analysis of inclination and eccentricity, it also proved possible to examine the effects of resonance on the along-track movement.[4] This was a new departure and the possibility of small errors cannot be entirely excluded. However, the nominal accuracy of the results is about 3 times better than in any previous resonance analysis. The value of the best lumped 14th-order harmonic from analysis of the mean motion of Cosmos 1603 was -20.7 ± 0.1, as compared with -20.8 ± 1.3 from analysis of the inclination (which was near 71°). Nominally the most accurate previous value (though for a different inclination) was -19.5 ± 0.5, from Meteor 5 rocket at 81°, and, for 15th order, -16.0 ± 0.2 from 1971-54A at 90° (all values $\times 10^{-9}$). So the nominal accuracy has been improved from a previous best of about 1.5% down to 0.5%.

I continued to be consulted in these university studies because I remained a member of the Editorial Board of the journal *Planetary and Space Science* until 1991 and was responsible for recommending most papers on orbit analysis for publication, after discussion with the authors if necessary. This work was a major element in my thirty years of cooperation with Sir David Bates, the editor of the journal.

A new area of university work has arisen out of the intention at the RAE to continue orbit analysis via a contract with a researcher in Academia after my retirement. At first the idea had been to make this arrangement with Bill Boulton at Coventry Polytechnic, but he left to take a post in Australia; and the contract was not finalized until 1990, with Clive Brookes, who then transferred from Aston to the University of Birmingham. He is working there in the Mathematics Department on the long-term analysis over several years of the orbits of Cosmos 373 and Cosmos 185 rocket. These satellites have orbital inclinations of 62.9° and 64.0° respectively, close to the critical inclination of 63.4°; they have been observed intensively over a number of years. The work should improve knowledge of the odd zonal harmonics in the gravity field.

Though the visual and other observations have been well used up to now, who can tell what the future will bring if UK science continues to decline? (One ironic answer is that, if it declined far enough, orbit analysis might be the only space research the country could afford; but so drastic a decline is unlikely.) Future orbit analysis will suffer from the absence of Hewitt camera observations, which have been available for all the satellites mentioned in the four preceding paragraphs. For resonance analysis, the lack of Hewitt camera observations can be largely compensated for by working only with slow resonances, of satellites with low drag, which accumulate large perturbations. It is worth remembering that the analysis of the five-year resonance of 1971-54A (see Fig. 6.6) did not use Hewitt camera observations and yet gave the most accurate of all the values of lumped harmonics (prior to Cosmos 1603).

The absence of the Hewitt cameras will, however, greatly hamper the study of atmospheric rotation rates from orbital changes. This is not as bad as it might be because, as already mentioned, this technique was gradually becoming outmoded; and the existing picture of the atmospheric rotation rate (Fig. 6.12) is fairly adequate for removing orbital perturbations, though not at heights above 600 km.

The high reputation of orbit analysis in the 1960s had declined during the 1970s and 1980s, when the techniques were (ironically) much better: but it was not entirely forgotten, for I was awarded the Nordberg Medal of COSPAR in 1990. The medal is given for 'distinguished contributions in the application of space science in a field covered by COSPAR' and had only been awarded once before, in 1988. This recognition of orbit analysis, when so many other areas of space science could have been chosen, was a pleasant ending to my long association with COSPAR. There are also two postscripts to my even longer association with the RAE: in January 1991 I became a consultant to the RAE and the University of Birmingham in an extension of the MoD contract with Clive Brookes; and in April 1991 the RAE changed its name to Aerospace Division, Defence Research Agency.

Although obliged to retire from the RAE at 60 (or 60½ to be precise), I was still regarded as quite young by the Royal Society, the average age of Fellows being 65. I remained as chairman of the National Committee for the History of Science, Medicine and Technology, until it also was 'dissolved' at the end of 1989. Unlike the other twenty-eight National Committees, however, it was a grant-giving body, and the grants in history of science still had to be allocated. So again a phoenix arose from the ashes, and a new History of Science Grants Committee, with the same membership as the National Committee, emerged in 1990, shorn of international duties but with added responsibilities in overseeing the general health of history of science research and teaching in the UK. This was the only Committee to remain inviolate (in membership, if not in name), and I have so far continued as chairman, sitting still while the work is admirably done by Sheila Edwards, the Royal Society's Librarian. The Committee's grants are quite small, but they are very helpful to historians of science needing to travel to libraries and archives for research.

Another committee-task at the Royal Society with a visible and practical outcome was as a member of a small 'Archives Working Group' headed by Professor Patricia Clarke. Several varied meetings with archival people and products, ranging from the Keeper of the Public Records to the British Library's latest computerized catalogues, convinced us of the great virtues of ink on paper, and the many problems, costs and possible impermanence of technology that is currently trendy but in thirty years may either self-destruct or be unreadable (like our 300000 satellite observation cards). The Archives Working Group recommended that the existing old papers and books should be kept as they were, and that a new air-conditioned strong-room be built in the basement of the Royal Society to replace the existing dusty shelves. The recommendation was accepted, the money was found and in 1989 the new Archives Room was built. In 1988 I rejoined the Royal Society's Library Committee – one of the oldest of committees, first appointed in 1678.

The *Philosophical Transactions* and *Proceedings* of the Royal Society are well known to scientists, but another journal with the rather dry title *Notes and Records of the Royal Society of London* is less known (except to Fellows, who all receive a copy). The subject matter of *Notes and Records* is the history of the Royal Society and its membership, and of the science, medicine and technology associated with it. In 1988 the Royal Society made changes in the editorial arrangements for all its journals, and I was asked to become editor of *Notes and Records*, in succession to the previous joint editors, Professor R. V. Jones and Sir William Paton, who had guided the journal for nearly twenty years. Being an editor is a never-ending task: however, the subject interested me, and the task might not be too onerous, as the journal appears only twice a year, with a annual tally of about 300 pages. In fact there has been more work than expected, because of time spent deciding whether papers are suitable or unsuitable in subject matter. I also knowingly created more work by starting book reviews. The first issue under my editorship was in January 1990, and so far all has gone smoothly, thanks to good deeds in the publications office at the Royal Society.

Another practical initiative in the history of science in the late 1980s was the transfer of the Contemporary Scientific Archives Centre at Oxford to a new home at the University of Bath, where it was renamed the National Cataloguing Unit for the Archives of Contemporary Scientists, under the direction of Professor Angus Buchanan. The original Centre at Oxford was set up in 1973 to take in the papers of eminent scientists and engineers recently deceased, to catalogue them, and to find a suitable archive where they would be accepted for deposit. Previously, such papers were often lost without trace. The Royal Society has provided more than half the finance from its history of science grant. I was much involved with the decision to transfer to Bath, and the transfer has so far proved most successful, with two archivists now at work and good hopes of maintaining the finance at its present level. The Vice-Chancellor of the University of Bath, Professor Rodney Quayle, took a keen interest in the Unit and acted as chairman of the Advisory Committee: I have been a rather inactive member of this body (which a new member once characterized as the 'Committee of the Freshly Dead').

Erasmus Darwin, though not so freshly dead, has continued to fascinate me in these recent years, particularly through his creation and use of words. I found more than 50 words in his translations of Linnaeus that were earlier than any usage recorded in the *Oxford English Dictionary*; and, after making a concordance to his poem *The Botanic Garden*, I found 128 such 'early words' there. Some of these were put on show in an article for *Notes and Records* in 1988 (before I was editor!):

The most fertile areas in *The Botanic Garden* for the growth of new words were its pastoral scenery, science and sheer luxury. Why not start in a flower garden on a summer day? There you recline on pillowy moss beside a pansied bank, watching butterflies jostle in aurelian splendour and savouring the scents of a hundred flowers borne on the wafting wing of a susurrant breeze too gentle to excite the

Eolian lyre standing in the open uncurtained window, or to rimple the smooth surface of the nearby lake with its pine-capt island – and this convoluted sentence has ten of Darwin's words (pillowy, pansied, aurelian, wafting, susurrant, Eolian, uncurtained, rimple, pine-capt, convoluted)....[5]

– and there were more silly sentences to display others. The 200th anniversary of the publication of Darwin's poem *The Loves of the Plants*, which was the second part (but the first published) of *The Botanic Garden*, was celebrated in April 1989 with a well-illustrated article in *New Scientist* entitled 'Chronicle of the lustful plants',[6] and I also wrote a chapter on Darwin for the *Dictionary of Literary Biography*.[7] An article on William Blake's poem 'The Tyger', a listing of the 'new words' used by Darwin, for *Notes and Queries*, and a number of book reviews, for the *Times Literary Supplement*, *Nature* and other journals, were among further literary trifles.

Questionings

This book has given my view of the research based on analysis of satellite orbits at the Royal Aircraft Establishment, Farnborough. That story now is ended; but perhaps, like the visiting lecturer at the local Society, I should try to answer some pointed questions.

The first question might well be, 'Was it all worth while?' My selfish answer is 'yes'. I was lucky in being born at the right time to be able to enter with ease a system – the scientific civil service – that was congenial and, as it turned out, allowed full scope for my interests and talents. Had I been born earlier, the war would have disrupted my educational progress. Had I been born later, I would not have been 'in place' ready to seize on the orbits of the first satellites and to prise out some of the rich store of information hidden in them. Also, a greater part of my career would then have been darkened by the decline of British science, and of staff numbers at the RAE. As it was, many of the human services I relied on at the RAE (typing, dictation, tracing, printing and libraries) endured almost intact to the end; and the computing facilities improved in the last decade. I was also lucky in being able to work on a subject that had always fascinated me – orbits evolving under the inexorable rule of Newton's laws – and occasionally to improve knowledge of the Earth and its atmosphere, thus conforming to the aims of 'the Royal Society of London for Improving Natural Knowledge', to give its full title.

The second awkward question might be: 'If orbit analysis gave information about the Earth and atmosphere more cheaply than other techniques of space research, why the faltering in the 1970s and 1980s?' A glib answer would be: because orbit analysis was not taken up in universities until the mid 1970s, and then only in one (or perhaps $1\frac{1}{2}$, one at Aston and a half at Leicester); whereas other and more expensive forms of space research attracted wider interest in UK universities, and inevitably acquired nearly all the funds. The well-heeled physicists and astronomers who secured this money would ('democratically') sit on the committees making future awards (each going out of the room when his turn came), and they also often chose the new members of the committee, who tended to be friends and colleagues in their own disciplines. Thus, by classic Darwinian natural selection, the survival of the financially fattest was assured: they, and the well-tailored administrators, looked down on the orbit analysts in their cut-price clothes.

The money needed to launch one spacecraft could have funded orbit analysis for twenty years and thereby produced more research results (though in limited fields, it is true). But the new spacecraft would be preferred for several reasons: (a) because it was identifiable as a 'prestigious project' to administrators; (b) because committee members might have experiments aboard; (c) because no industrial company could make profits from orbit analysis; and (d) because the orbit analysts were mathematicians, whereas the committee members were mostly astronomers and physicists. There was a fifth reason, too, arising from the inevitable subject divisions between the SERC, which covered space and astronomy, and the Natural Environment Research Council, which dealt with the Earth and its environment. Orbit analysis uses space techniques, but the results apply to the Earth and its environment; so the work fell neatly into the chasm between the two Research Councils, a situation that prompted several ineffectual protests. I met the chairmen of the two Councils in 1984 to present to them the problems over the support of geodesy. Both of them (Sir John Kingman and Sir Hermann Bondi) appreciated and agreed with my thesis, and tried to put things right, but with limited success because all new initiatives had to go through the appropriate committees in both Councils, with their understandable biases.

The 'reasons' above are valid: but orbit analysis would have fared much better if I had been a different person, not a silent scholar absorbed in the detail but an entrepreneur keen to promote the subject. And that I could never be, because I found meetings so boring (though useful for copying out satellite observations onto the proper forms, or recopying messy manuscripts).

So much for the past: are there any inspiring hopes for the future of orbit analysis? What is it all in aid of? This needs two answers, scientific and worldly. Scientifically, the broad-brush pictures of the upper-atmosphere density in the 1960s were timely and informative, but orbit analysis is no longer the prime method because it gives densities averaged widely in latitude and local time. However, orbit analysis is still – and will continue to be – the best method for monitoring the vagaries of the semi-annual variation, which has not yet been modelled or understood satisfactorily. Evaluating average atmospheric rotation rates from orbit analysis remains a valid method: it does not show up local effects, but does give the general picture, which still needs much improvement, especially at heights above 500 km, where accurate orbits determined from laser observations should give the best results. In the evaluation of the Earth's gravity field, orbit analysis will (unless entirely new methods arise) remain an important technique, pre-eminent for resonances, very effective for nearly resonant orbits and useful to varying extents for all others.

The new knowledge of the Earth uncovered by these researches is sought primarily for its own sake, just as astronomy helps in understanding the heavens with very few practical applications. But the evaluation of the Earth's gravitational field, the main focus of research in the 1980s, does have important practical applications in oceanography. Ocean tides and currents, and the effects of winds on the ocean, all produce departures of the ocean surface from the mean-sea-level geoid surface – and all are important in predicting the magnitude of the future 'greenhouse effect'. The geoid surface needs to be known to an accuracy of better than 10 cm if these departures from the average are to be measured with adequate accuracy from altimeter satellites, because the departures from the average are often only of order 10 cm. The orbit of the altimeter satellite also needs to be known to 10 cm in radial distance, which in turn calls for an accurate model of the gravity field, so that the orbital perturbations can be correctly modelled. This is a second and separate requirement that is also equivalent to an accuracy of about 10 cm in geoid height. Such accuracy is not yet available. So there is a pressing need for improved evaluation of the gravity field.

The scientific future may be quite bright, therefore; but the worldly prospects, at least in the UK, are not. With the family silver sold off, much of the old manufacturing base eroded, and less of the oil revenues that limited the decline in the 1980s, luxuries like science and culture may be squeezed out as real standards of life decline. Perhaps, as orbit analysis is cheap, it might slip through the meshes of a financial net that catches larger projects? Possibly: but, to revert to my favourite example, I do not expect to see in the next twenty years a 32-year-old government research scientist earning in a year 20% more than the price of a three-bedroomed house, just for doing congenial research with no administrative responsibilities. And that is the real measure of what has been lost, in worldly terms.

I was not worldly: few scientists are; but was I really a scientist? Most research scientists pursue research full-time, whereas a fair proportion (perhaps 30%) of my total 'intellectual time', at work and at home, was devoted to writing books on varied subjects, literary, historical – and scientific. The habit began in the mid 1950s when the guided missiles occupied working hours only (partly because of the security requirements). So my intellectual interests became divided. The two halves have greatly helped each other. My scientific career enhanced the credibility of books such as the *Shelley*, which dwelt on his scientific interests and imagery. My literary-historical books, implying a wider view than that of most scientists, led to unusually many invitations to give special lectures, which (strange as it may seem) enhance scientific reputations. The best example is the Bakerian Lecture that I gave in 1974: the next lecturer, in 1975, was Michael Atiyah, and he was followed in 1977 by George Porter – two future Presidents of the Royal Society.

Most of my books have been 'scholarly', with footnotes to validate the text, and I have never felt any essential intellectual difference between the writing of scientific papers and the literary-historical-biographical books. A somewhat different style of writing is needed, but no more different than between one literary genre and another – between a passionate poem and a factual footnote to one of Erasmus Darwin's letters.

The chief differences between the scientific and literary worlds arise because (apart from a few hiccups) science continually *advances*, whereas literary scholarship may go backwards in the search for novelty, and new creative writing, though novel, is not necessarily better: a bright school-child can outdo Newton in mathematics; but who can outdo Shakespeare? Because science advances, scientists must be up-to-date; literary pundits can be more leisurely. Literary scholars need to have much better memories, carrying in their heads the plots of a thousand novels. With science, memory is not as important as being receptive to new ideas: memories may need to be erased if a new paradigm burgeons and prevails within a year, whereas a new literary orthodoxy usually takes decades. Another difference is in the presentation of lectures: literary scholars usually read the written text of a paper; scientists usually show numerous slides or 'vugraphs', and just talk about them. My own method was, appropriately, half-way between these extremes. For any special lecture I dictated beforehand a spoken version, which the shorthand typist produced in outsize 'Orator' type. I could revert to reading this if I forgot what I intended to say, as happened quite often.

I do not see any essential difference in creativity between science and literature: some branches are more creative than others, in both subjects. You could say that finding better values for harmonics is not creative, but creativity is needed in devising the techniques of evaluation. Similarly, literary criticism could be called second-hand, or even parasitic, but creative insights do often occur. At the other extreme are the speculative astronomers and imaginative poets or novelists, who all tend to create new world-systems, usually imaginary, from their fertile imaginations. (Fred Hoyle, eminent among imaginative astronomers, is well known too for science fiction.)

The same chorus-line of 'science advances' emerges from my other cross-cultural experience, in travelling from science to history of science. When I became Chairman of the National Committee for the History of Science, Medicine and Technology, I was soon involved with the history of science community, academic and amateur, whose grants the Committee had to consider. I found the community divided in its attitudes. Most academic historians of science stress the setting of science within the social milieu of its era, and look at the social pressures that have moulded that science. At its most extreme, this turns history of science into social history. At its most fruitful, in the history of medicine, the result has been a new view of the subject from the patient's standpoint: the work of Roy Porter at the Wellcome Institute for the History of Medicine has been important in effecting this change. The other group in the history of science community consists of scientists interested in the history of their subject, and a minority of the academics. The scientists had originally become scientists because they liked the idea of trying to advance science – 'improving natural knowledge...'. They tend to see the past advancing towards the present, and to look with approval on past scientists who came close to modern ideas. The idea of progress is still alive, deep in the subconscious of every true scientist. This faith that things are getting better tends to be scoffed at by the 'social historian' group as part of the 'Whig view of history'. They are right to be cynical, because the lessons of history are never learnt, and wars continue to be fomented; they are right too for art, literature and music, which go in cycles; but they are wrong for science, which continually advances as natural knowledge is improved.

All this intellectual chattering (so unlike me) leaves me with the uneasy feeling that, even though not pretending to be a fully-fledged scientist, I am still misleading everyone by implying that I am an intellectual, which is not true, because I never consciously think, but just leap to the obvious conclusion. I do not have much patience with creeds; but most things in life are explained by biological evolution as expounded in the strong form (with organic happiness) by Erasmus Darwin, and more mildly and scientifically by his grandson Charles. So I have always been a devotee of physical activity, playing tennis at least twice a week and lately five or six times, with cross-country running and appropriate amounts of other healthy sports. Also I do still retain the old-fashioned belief that science brings the power to make our lives better to live, not just through the electric light and wonder drugs, but also through the filtration into practical life of advances in fundamental knowledge in a thousand-and-one separate specialized areas. Today people tend to complain about science, blaming it for everything from vivisection to nuclear weapons. I hope that attitude will soon reverse again: science will help us to a better future if we are wise enough to want that.

References

Chapter 1 (pp. 4–20)

1 D. G. King-Hele and D. M. C. Gilmore, RAE Technical Notes GW 332, 355, 376, 417 and 438 (1954–57).
2 D. G. King-Hele and D. M. C. Gilmore, 'Preliminary assessment of an earth satellite reconnaissance vehicle', RAE Technical Note GW 393 (January 1956).
3 D. G. King-Hele and D. M. C. Walker, *Vistas in Astronomy*, **30**, 269–289 (1987).
4 D. G. King-Hele and D. M. C. Gilmore, RAE Technical Note GW 455 (May 1957).
5 R. H. Gooding, RAE Technical Report 88068 (1988).

Chapter 2 (pp. 21–44)

1 Staff of RAE, *Nature*, **180**, 937–941 (1957).
2 R. A. Minzner and W. S. Ripley, 'The ARDC model atmosphere 1956', US Air Force Surveys in Geophysics No. 86 (1956).
3 *CIRA 1972*: COSPAR International Reference Atmosphere 1972 (Akademie, Berlin, 1972).
4 *Nature*, **181**, 738 (1958).
5 D. G. King-Hele and D. C. M. Leslie, *Nature*, **181**, 1761–1763 (1958).
6 R. H. Merson and D. G. King-Hele, *Nature*, **182**, 640–641 (1958).
7 R. H. Merson, D. G. King-Hele and R. N. A. Plimmer, *Nature*, **183**, 239–240 (1959).
8 D. G. King-Hele and D. M. C. Walker, *Nature*, **182**, 426–427 (1958).

Chapter 3 (pp. 45–73)

1 D. G. King-Hele and D. M. C. Walker, *J. Brit. Interplan. Soc.*, **17**, 2–14 (1959).
2 D. G. King-Hele and D. M. C. Walker, *Nature*, **183**, 527–529 (1959).
3 L. G. Jacchia, *Nature*, **183**, 526–527 (1959).
4 L. G. Jacchia, Smithsonian Astrophys. Obs. Spec. Rpt 100 (1961).
5 D. G. King-Hele, *Nature*, **183**, 1224–1227 (1959).
6 D. G. King-Hele, *Proceedings of the Tenth International Astronautical Congress, 1959* (Springer Verlag, Vienna, 1960) pp. 1–20.
7 D. G. King-Hele and D. M. C. Walker, *Nature*, **186**, 928–931 (1960).

8 D. G. King-Hele and D. M. C. Walker, *Annales de Géophysique*, **17**, 162–171 (1961).
9 D. G. King-Hele, G. E. Cook and D. M. C. Walker, 'The contraction of satellite orbits under the influence of air drag. Part I: with spherically symmetrical atmosphere', RAE Technical Note GW 533 (1959); and *Proc. Roy. Soc.*, **A257**, 224–249 (1960).
10 D. G. King-Hele, G. E. Cook and D. M. C. Walker, RAE Technical Note GW 565 (1960); and *Proc. Roy. Soc.*, **A264**, 88–121 (1961).
11 D. G. King-Hele, *Proc. Roy. Soc.*, **A253**, 529–538 (1959).
12 D. G. King-Hele and R. H. Merson, *Nature*, **183**, 881–882 (1959).
13 D. G. King-Hele, *Nature*, **187**, 490–491 (1960).
14 D. G. King-Hele, *Geophys. J. Roy. Astronom. Soc.*, **4**, 3–16 (1961).

Chapter 4 (pp. 74–126)

1 D. G. King-Hele and D. M. C. Walker, *Space Research II* (North-Holland, Amsterdam, 1961), pp. 918–957.
2 H. Hiller, *Astronautica Acta*, **8**, 82–105 (1962).
3 V. A. Yegorov, 'Problems on the dynamics of flight to the Moon', *The Russian Literature of Satellites*, Part I (New York, 1958).
4 R. H. Merson, *Geophys. J. Roy. Astronom. Soc.*, **4**, 17–48 (1961).
5 D. G. King-Hele, *Satellites and Scientific Research* (Routledge, 1960), pp. 64–5.
6 D. G. King-Hele and J. M. Rees, *J. Atmos. Terr. Phys.*, **25**, 495–506 (1963).
7 D. G. King-Hele and E. Quinn, *J. Atmos. Terr. Phys.*, **27**, 197–209 (1965).
8 D. G. King-Hele, *Poems and Trixies* (Mitre Press, London, 1972), p. 18 (and p. 9).
9 D. G. King-Hele and D. M. C. Walker, *Nature*, **220**, 775 (1968), and *Planet. Space Sci.*, **17**, 985–997, 1539–1556 and 2027–9 (1969).
10 D. G. King-Hele, *J. Brit. Interplan. Soc.*, **19**, 374–382 (1964).
11 D. G. King-Hele and J. Hingston, *Planet. Space Sci.*, **16**, 937–949 (1968).
12 D. G. King-Hele and D. M. C. Walker, *Nature*, **219**, 715–716 (1968); *Planet. Space Sci.*, **17**, 197–215 (1969).
13 G. E. Cook and D. W. Scott, *Planet. Space Sci.*, **15**, 1933–1956 (1967).
14 G. E. Cook, *Annales de Géophysique*, **25**, 451–469 (1969).
15 D. G. King-Hele and J. M. Rees, *Proc. Roy. Soc.*, **A270**, 562–587 (1962).
16 D. G. King-Hele, *Contemporary Physics*, **2**, 253–267 (1961).
17 D. G. King-Hele, *Planet. Space Sci.*, **12**, 835–853 (1964).
18 D. G. King-Hele and D. W. Scott, *Planet. Space Sci.*, **14**, 1339–1365 (1966).
19 D. G. King-Hele and D. W. Scott, *Nature*, **213**, 1110 (1967).
20 D. G. King-Hele and D. W. Scott, *Planet. Space Sci.*, **15**, 1913–1931 (1967).
21 D. G. King-Hele, D. W. Scott and D. M. C. Walker, *Planet. Space Sci.*, **18**, 1433–1445 (1970).
22 D. G. King-Hele, *Proc. Roy. Soc.*, **A267**, 541–557 (1962).
23 G. E. Cook and D. G. King-Hele, *Proc. Roy. Soc.*, **A275**, 357–390 (1963).
24 G. E. Cook and D. G. King-Hele, *Phil. Trans. Roy. Soc.*, **A259**, 33–67 (1965).
25 G. E. Cook, *Planet. Space Sci.*, **14**, 433–444 (1966).
26 G. E. Cook and D. G. King-Hele, *Proc. Roy. Soc.*, **A303**, 17–35 (1968).
27 D. G. King-Hele and D. W. Scott, *Planet. Space Sci.*, **17**, 217–232 (1969).
28 D. G. King-Hele, *Proc. Roy. Soc.*, **A294**, 261–272 (1966).
29 R. R. Allan, *Q. J. Mech. App. Math.*, **15**, 283–301 (1962).
30 R. R. Allan, *Planet. Space Sci.*, **11**, 1325–1334 (1963).

31 G. E. Cook, *Geophys. J. Roy. Astronom. Soc.*, **6**, 271–291 (1962).
32 G. E. Cook, *Planet. Space Sci.*, **11**, 797–815 (1963); **12**, 1009–1020 (1964).
33 H. Hiller, *Planet. Space Sci.*, **13**, 147–161 (1965); **13**, 1233–1247 (1965): **14**, 773–789 (1966).
34 E. G. C. Burt, *Proc. Roy. Soc.*, A**308**, 217–241 (1968).
35 D. G. King-Hele, G. E. Cook and D. W. Scott, *Planet. Space Sci.*, **13**, 1213–1232 (1965).
36 D. G. King-Hele, G. E. Cook and D. W. Scott, *Planet. Space Sci.*, **15**, 741–769 (1967).
37 D. G. King-Hele, G. E. Cook and D. W. Scott, *Planet. Space Sci.*, **17**, 629–644 (1969).
38 'A Discussion on orbital analysis', *Phil. Trans. Roy. Soc.*, A**262**, 1–202 (1967).
39 D. G. King-Hele, D. M. C. Walker and P. E. L. Neirinck, *Nature*, **221**, 130–2 (1969).
40 D. G. King-Hele, *Russell*, **16**, 21–26 (1975).

Chapter 5 (pp. 127–168)

 1 R. R. Allan, *Proc. Roy. Soc.*, A**288**, 60–68 (1962); *Planet. Space Sci.*, **15**, 53–76 (1967); **15**, 1829–1845 (1967); **21**, 205–225 (1973).
 2 R. H. Gooding, *Nature Phys. Sci.*, **231**, 168–9 (1971).
 3 D. G. King-Hele, *Nature Phys. Sci.*, **238**, 13 (1972); also *Space Research XIII* (Akademie, Berin, 1973), pp. 21–9.
 4 E. M. Gaposchkin *et al.*, Smithsonian Astrophys. Obs. Spec. Rpt 353 (1973).
 5 F. J. Lerch *et al.*, Goddard Space Flight Center Rpt X-921-74-145 (1974).
 6 D. G. King-Hele, *Q. J. Roy. Astronom. Soc.*, **13**, 374–395 (1972).
 7 D. G. King-Hele, D. M. C. Walker and R. H. Gooding, *Nature*, **249**, 748–750 (1974); also *Planet. Space Sci.*, **22**, 1349–1373 (1974).
 8 C. A. Wagner *et al.*, Goddard Space Flight Center Rpt X-921-76-20 (1976).
 9 D. G. King-Hele, *Planet. Space Sci.*, **22**, 509–524 (1974).
10 D. G. King-Hele and A. N. Winterbottom, *Planet. Space Sci.*, **22**, 1045–1057 (1974).
11 D. G. King-Hele, D. M. C. Walker and R. H. Gooding, *Planet. Space Sci.*, **23**, 229–246 (1975).
12 D. G. King-Hele, D. M. C. Walker and R. H. Gooding, *Planet. Space Sci.*, **27**, 1–18 (1979).
13 R. H. Merson, *Space Research XIII* (Akademie, Berlin, 1973), pp. 35–43.
14 D. M. C. Walker, *Planet. Space Sci.*, **25**, 337–342 (1977).
15 H. Hiller and D. G. King-Hele, *Planet. Space Sci.*, **25**, 513–520 (1977).
16 D. G. King-Hele, *Phil. Trans. Roy. Soc.*, A**278**, 67–109 (1975).
17 D. G. King-Hele and G. E. Cook, *Planet. Space Sci.*, **22**, 646–672 (1974).
18 D. G. King-Hele, *Science*, **192**, 1293–1300 (1976).
19 D. G. King-Hele, *Nature*, **223**, 325–6 (1971).
20 D. G. King-Hele and D. M. C. Walker, *Planet. Space Sci.*, **25**, 313–336 (1977).
21 D. G. King-Hele, *Planet. Space Sci.*, **24**, 1–16 (1976).
22 D. G. King-Hele and D. M. C. Walker, *Planet. Space Sci.*, **21**, 1081–1108 (1973).
23 D. G. King-Hele and D. M. C. Walker, *Planet. Space Sci.*, **19**, 297–311 and 1637–1651 (1971).
24 D. M. C. Walker, *Planet. Space Sci.*, **22**, 403–411 (1974).
25 D. M. C. Walker, *Planet. Space Sci.*, **26**, 291–309 (1978).

26 D. G. King-Hele, *Rev. Geophys. Space Phys.*, **16**, 733–740 (1978).
27 D. G. King-Hele, *Proc. Roy. Soc.*, A**330**, 467–494 (1972).
28 D. G. King-Hele and D. M. C. Walker, *Celestial Mechanics*, **5**, 41–54 (1972).
29 D. G. King-Hele and D. M. C. Walker, *Proc. Roy. Soc.*, A**350**, 281–298 (1976).
30 D. G. King-Hele, *J. Brit. Interplan. Soc.*, **31**, 181–196 (1978).
31 D. G. King-Hele, *J. Brit. Interplan. Soc.*, **28**, 783–796 (1975).
32 D. G. King-Hele, *Your Environment*, **2**, 6–10 (1971).
33 D. G. King-Hele, *The Observatory*, **95**, 1–12 (1975).

Chapter 6 (pp. 169–216)

1 *The RAE Table of Earth Satellites 1957–1986*, compiled by D. G. King-Hele, D. M. C. Walker, J. A. Pilkington, A. N. Winterbottom, H. Hiller and G. E. Perry (Macmillan Press, London, 1987).
2 D. G. King-Hele, C. J. Brookes and G. E. Cook, *Geophys. J. Roy. Astronom. Soc.*, **64**, 3–30 (1981).
3 C. Reigber *et al.*, *J. Geophys. Res.*, **90**, 9285–9299 (1985).
4 F. J. Lerch *et al.*, NASA Technical Memorandum 104555 (1992)
5 D. G. King-Hele, *Proc. Roy. Soc.*, A**374**, 327–350 (1980).
6 D. M. C. Walker, *Geophys. J. Roy. Astronom. Soc.*, **67**, 1–18 (1981).
7 F. J. Lerch *et al.*, *Marine Geodesy*, **5**, 145–187 (1981).
8 D. G. King-Hele and D. M. C. Walker, *Proc. Roy. Soc.*, A**379**, 247–288 (1982).
9 D. G. King-Hele and D. M. C. Walker, *Planet. Space Sci.*, **33**, 223–238 (1985).
10 D. G. King-Hele and D. M. C. Walker, *Planet. Space Sci.*, **35**, 79–90 (1987).
11 D. G. King-Hele and D. M. C. Walker, *Planet. Space Sci.*, **37**, 805–823 (1989).
12 D. G. King-Hele and D. M. C. Walker, *Planet. Space Sci.*, **30**, 411–425 (1982).
13 D. M. C. Walker, *Planet. Space Sci.*, **39**, 685–695 (1991).
14 D. G. King-Hele and D. M. C. Walker, *Planet. Space Sci.*, **38**, 407–409 (1990).
15 J. G. Marsh *et al.*, *J. Geophys. Res.*, **95**, 22043–22071 (1990).
16 D. M. C. Walker, *Planet. Space Sci.*, **33**, 1439–1449 (1985).
17 D. G. King-Hele and D. M. C. Walker, *Planet. Space Sci.*, **34**, 183–195 (1986).
18 D. M. C. Walker, *Proc. Roy. Soc.*, A**387**, 187–217 (1983).
19 D. G. King-Hele and D. M. C. Walker, *Planet. Space Sci.*, **35**, 937–946 (1987).
20 C. A. Wagner, *J. Geophys. Res.*, **79**, 3335–3341 (1974).
21 D. G. King-Hele, *Phil. Trans. Roy. Soc.*, A**296**, 597–637 (1980).
22 D. M. C. Walker, *Planet. Space Sci.*, **28**, 1059–1072 (1980).
23 A. N. Winterbottom, M. R. Suttie and D. G. King-Hele, *Planet. Space Sci.*, **36**, 449–458 (1988).
24 D. M. C. Walker, *Planet. Space Sci.*, **32**, 717–725 (1984).
25 D. M. C. Walker, *Planet. Space Sci.*, **37**, 351–362 (1989).
26 D. M. C. Walker, *Phil. Trans. Roy. Soc.*, A**292**, 473–512 (1979).
27 H. Hiller, *Planet. Space Sci.*, **27**, 1247–1267 (1979).
28 D. G. King-Hele and D. M. C. Walker, *Planet. Space Sci.*, **31**, 509–535 (1983).
29 T. Fuller-Rowell and D. Rees, *J. Atmos. Sci.*, **37**, 2545–2567 (1980).

30 D. G. King-Hele and A. N. Winterbottom, *Planet. Space Sci.*, **33**, 1125–1143 (1985).
31 D. G. King-Hele and D. M. C. Walker, *Planet. Space Sci.*, **36**, 1085–1093 (1988).
32 H. Hiller and D. G. King-Hele, *Planet. Space Sci.*, **29**, 35–45 (1981).
33 H. Hiller, *Planet. Space Sci.*, **29**, 574–588 (1981).
34 A. N. Winterbottom and D. G. King-Hele, *Planet. Space Sci.*, **32**, 1–16 (1984).
35 D. G. King-Hele, *Proc. Roy. Soc.*, **A393**, 235–256 (1984).
36 G. G. Swinerd and W. J. Boulton, *Proc. Roy. Soc.*, **A383**, 127–145 (1982); **A386**, 55–75 (1983); **A389**, 153–170 (1983).
37 W. J. Boulton, *Proc. Roy. Soc.*, **A389**, 349–367 (1983); **A389**, 433–444 (1983); **A391**, 201–214 (1984).
38 D. G. King-Hele and D. M. C. Walker, *Proc. Roy. Soc.*, **A411**, 1–17 (1987).
39 D. G. King-Hele and D. M. C. Walker, *Proc. Roy. Soc.*, **A411**, 19–33 (1987).
40 D. G. King-Hele and D. M. C. Walker, *Proc. Roy. Soc.*, **A414**, 271–295 (1987).
41 D. G. King-Hele and D. M. C. Walker, *Acta Astronaut.*, **18**, 123–131 (1988).
42 D. G. King-Hele, *Q. J. Roy. Astronom. Soc.*, **26**, 237–261 (1985).
43 D. G. King-Hele, *Notes Rec. Roy. Soc. Lond.*, **42**, 3 (1988).
44 D. G. King-Hele and D. M. C. Walker, *Vistas in Astronomy*, **30**, 269–289 (1987).

Chapter 7 (pp. 217–229)

1 *The RAE Table of Earth Satellites 1957–1989*, compiled by D. G. King-Hele, D. M. C. Walker, A. N. Winterbottom, J. A. Pilkington, H. Hiller and G. E. Perry (Royal Aerospace Establishment, Farnborough, 1990).
2 A. N. Winterbottom, *Planet. Space Sci.*, **38**, 1409–1420 (1990).
3 R. H. Gooding and D. G. King-Hele, *Proc. Roy. Soc.*, **A422**, 241–259 (1989).
4 M. S. Gilthorpe, P. Moore and A. N. Winterbottom, *Planet. Space Sci.*, **38**, 1147–1159 (1990).
5 D. G. King-Hele, *Notes Rec. Roy. Soc. Lond.*, **42**, 149–180 (1988).
6 D. G. King-Hele, *New Scientist*, **122** No. 1661, 57–61 (1989).
7 *Dictionary of Literary Biography*, Vol. 93 (Gale, Detroit, 1990), pp. 148–159.

Index

(The international designation is given after each satellite name)